I0419230

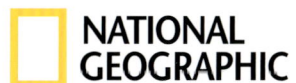

Photo
Ark
Babies

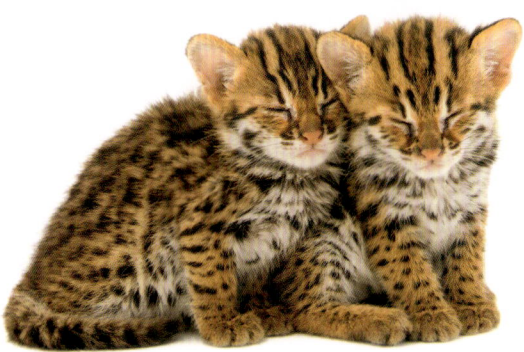

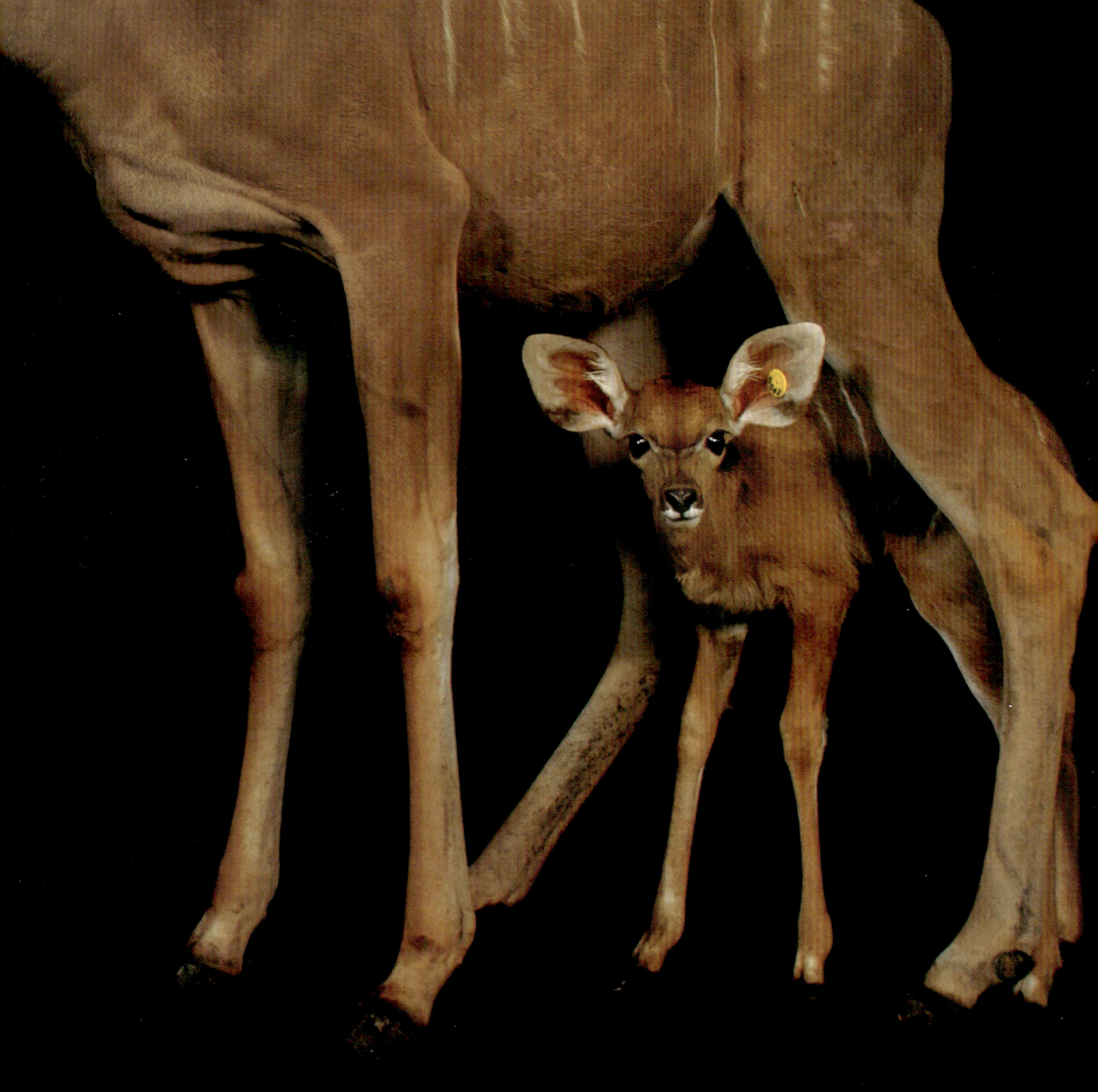

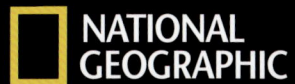

Photo Ark Babies

GROWING UP IN THE ANIMAL WORLD

JOEL SARTORE

NATIONAL GEOGRAPHIC
WASHINGTON, D.C.

FOR MY GRANDCHILDREN,
EZRA AND CASSIDY.
MAY THEY DELIGHT
IN ALL ANIMALS
AS MUCH AS I HAVE.

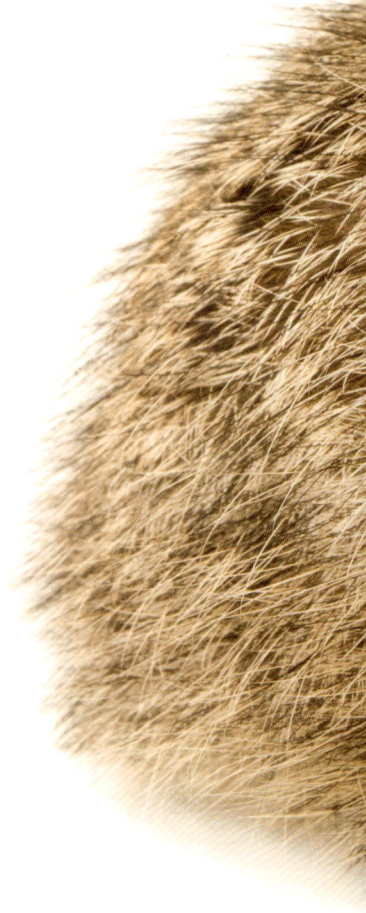

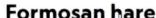

Formosan hare

Lepus sinensis formosus (LC)

Female hares give birth in burrows
to as many as three babies. Unlike rabbits,
whose young are blind and helpless
at birth, hares are born with fur, eyes open,
and are able to move around right away.

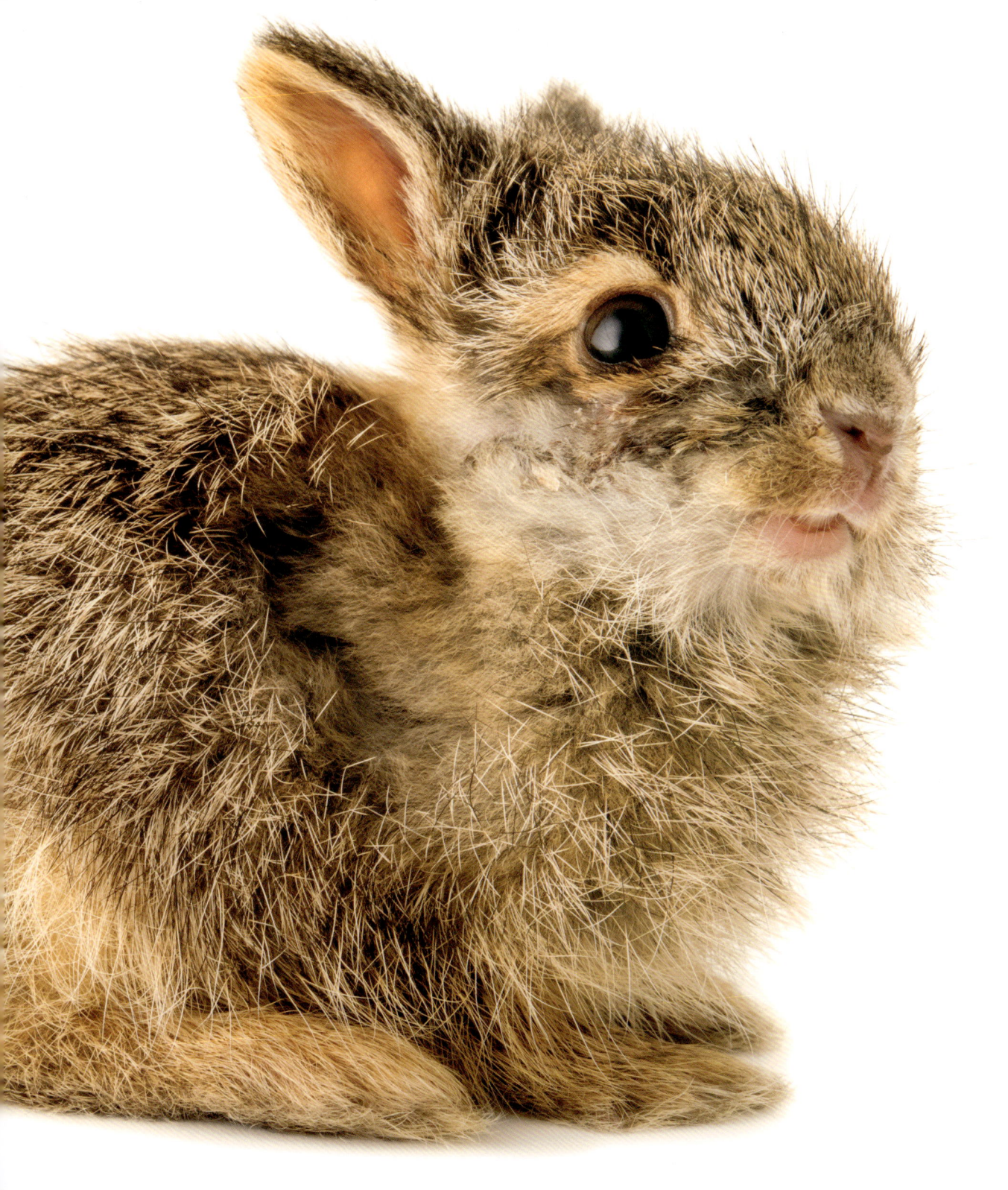

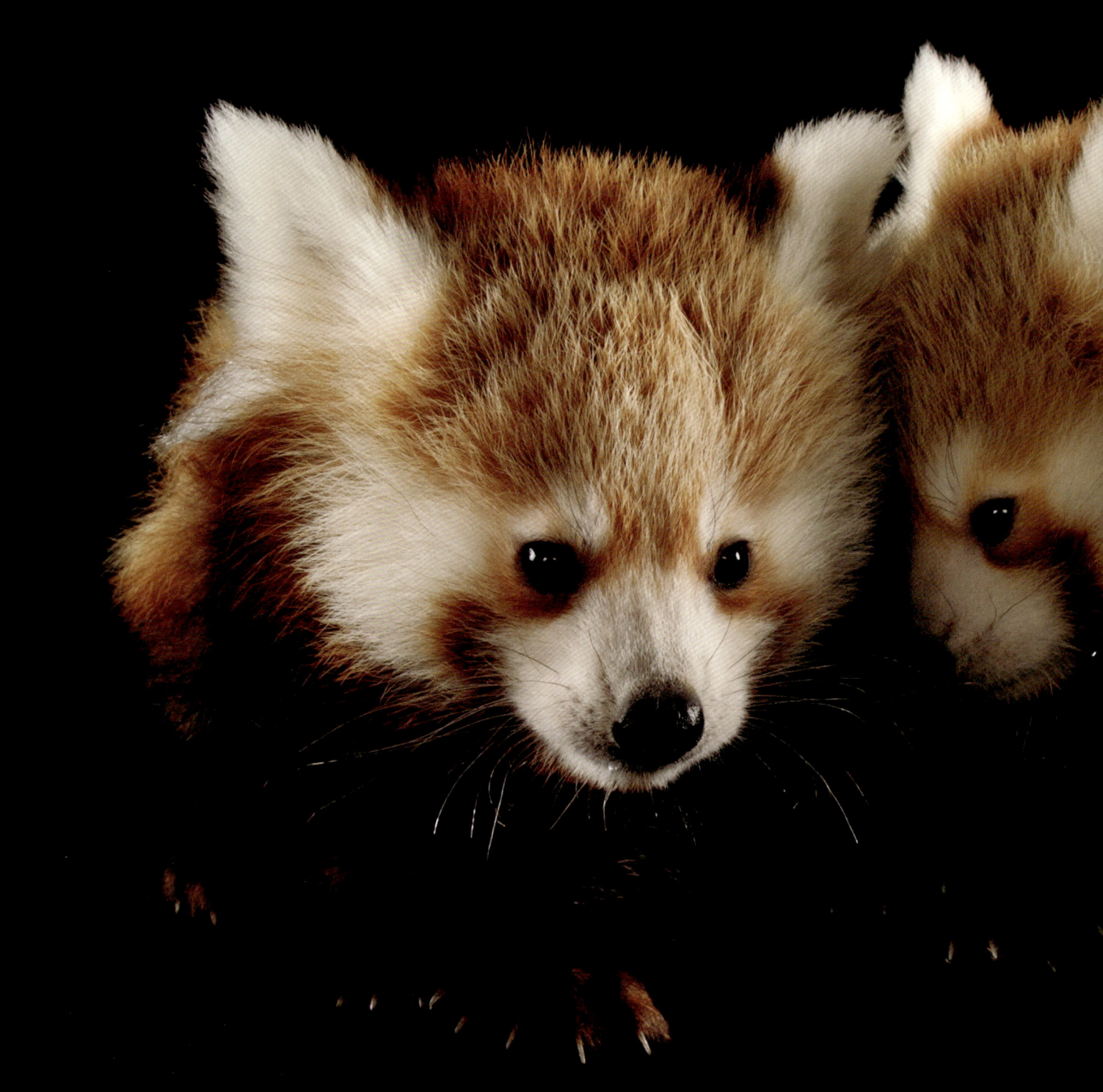

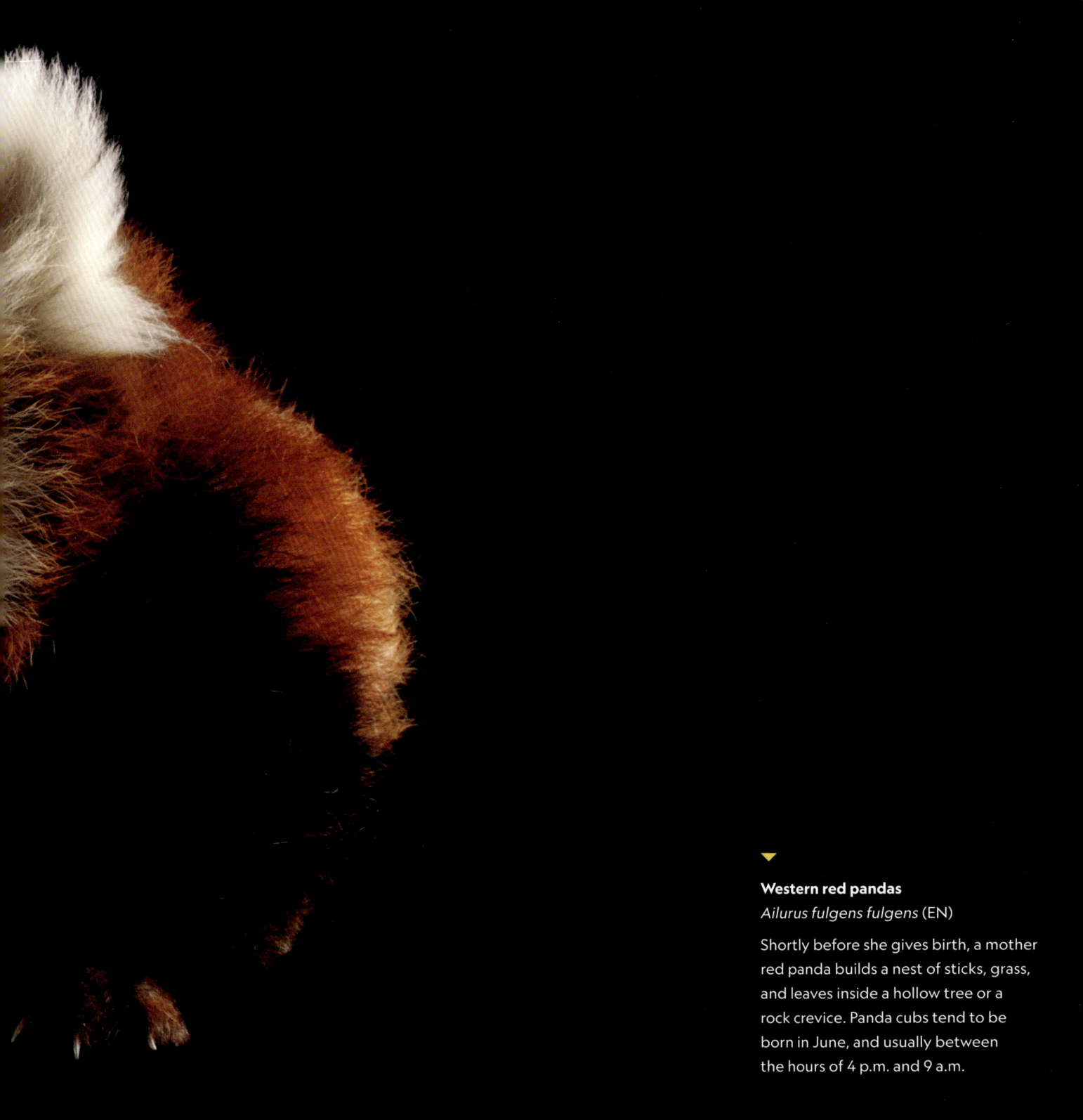

Western red pandas
Ailurus fulgens fulgens (EN)

Shortly before she gives birth, a mother red panda builds a nest of sticks, grass, and leaves inside a hollow tree or a rock crevice. Panda cubs tend to be born in June, and usually between the hours of 4 p.m. and 9 a.m.

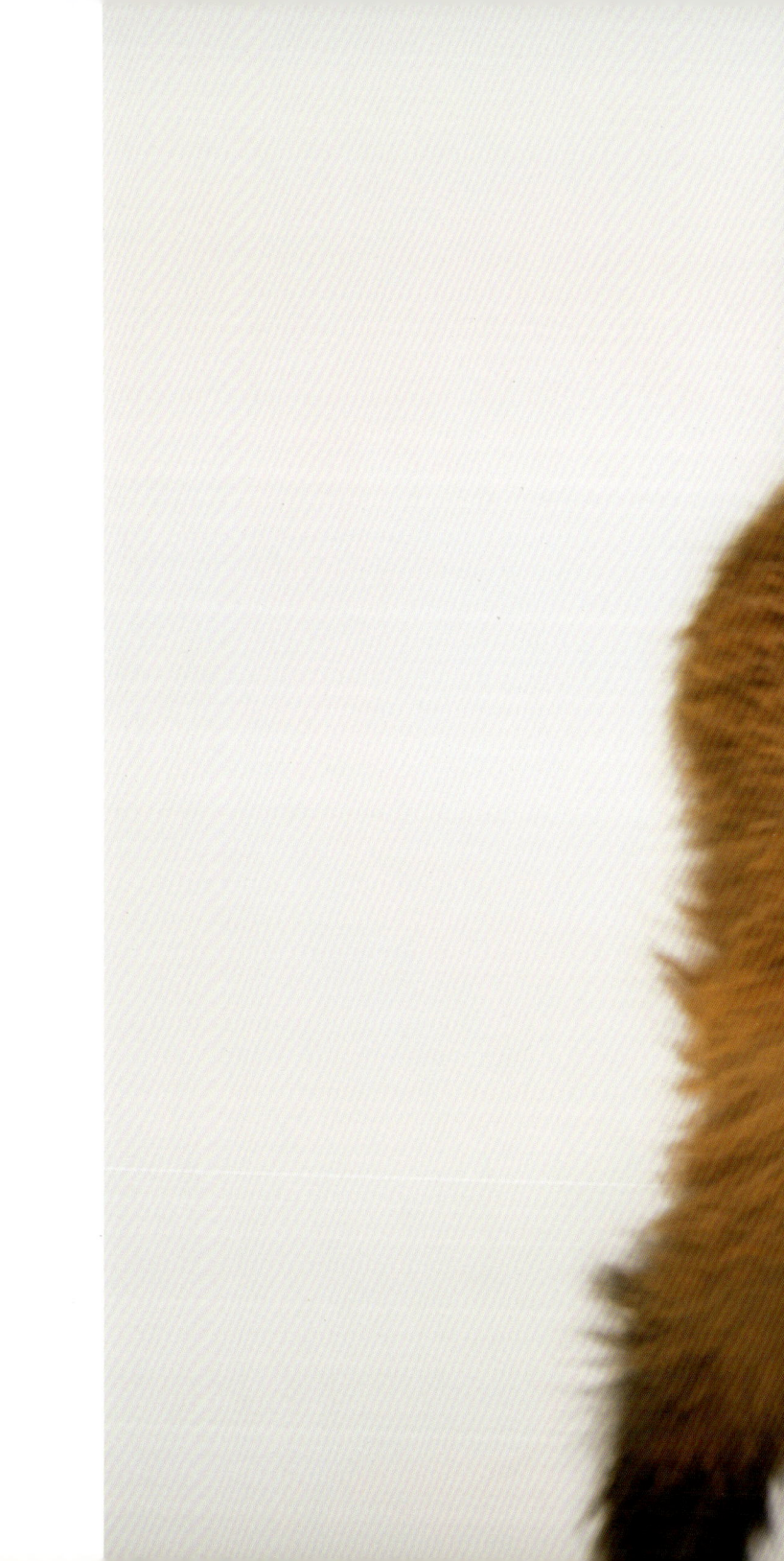

Red-flanked duiker

Cephalophus rufilatus (LC)

This small, rust-colored antelope of
the savannas and forests of Central and
West Africa hides amid grasses and
shrubs as soon as it's born, peeking
out only to nurse from its mother.

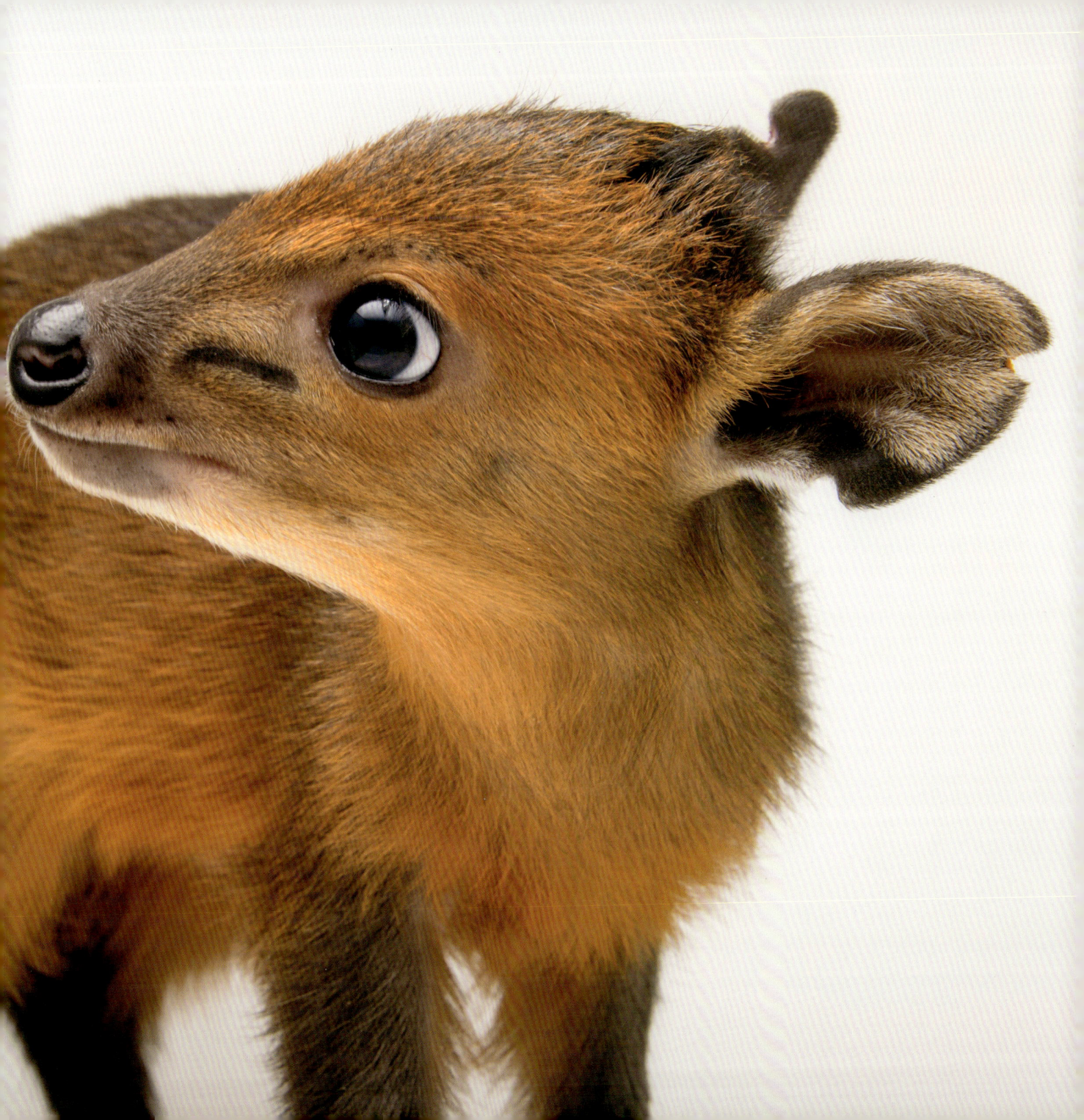

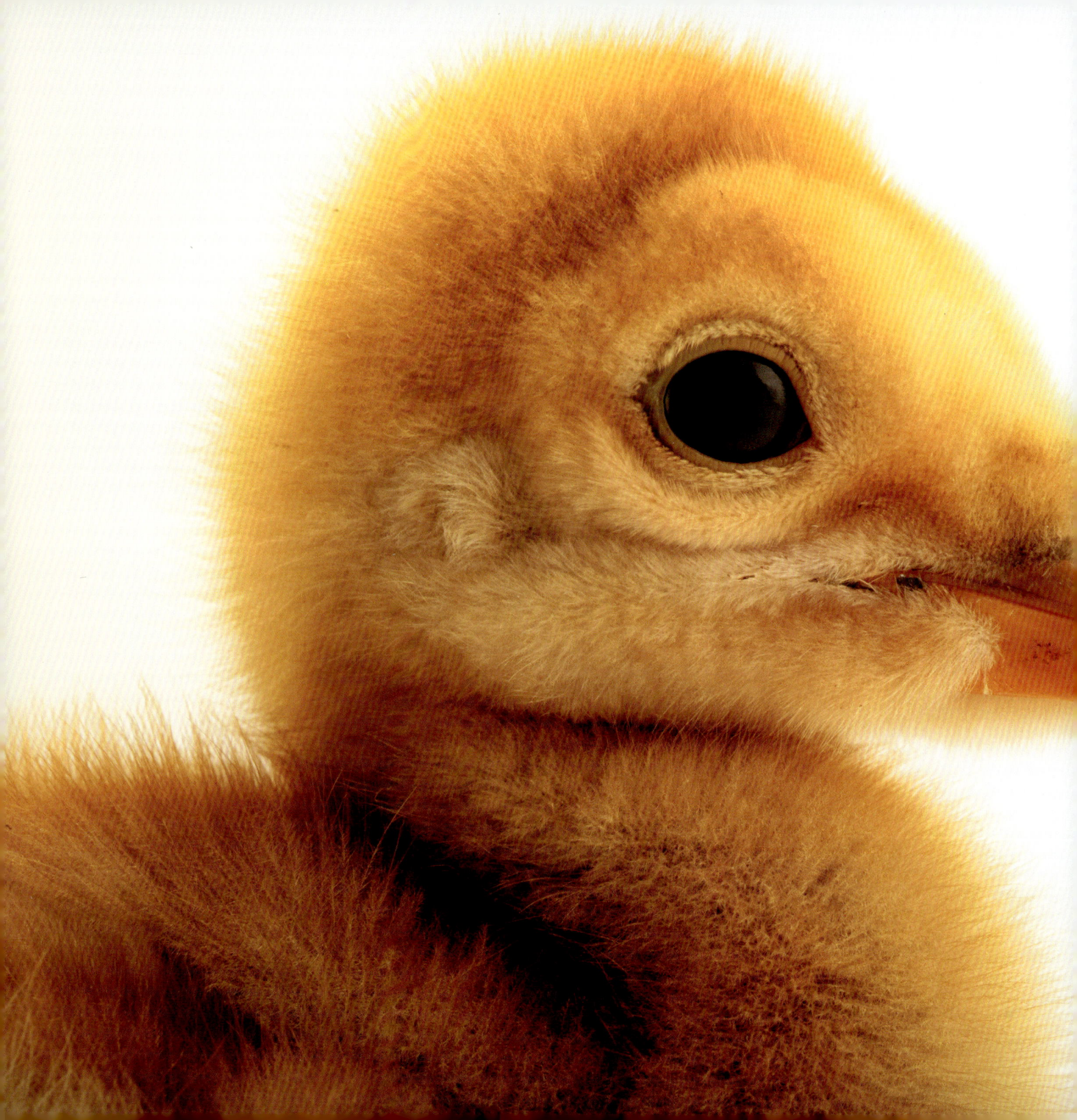

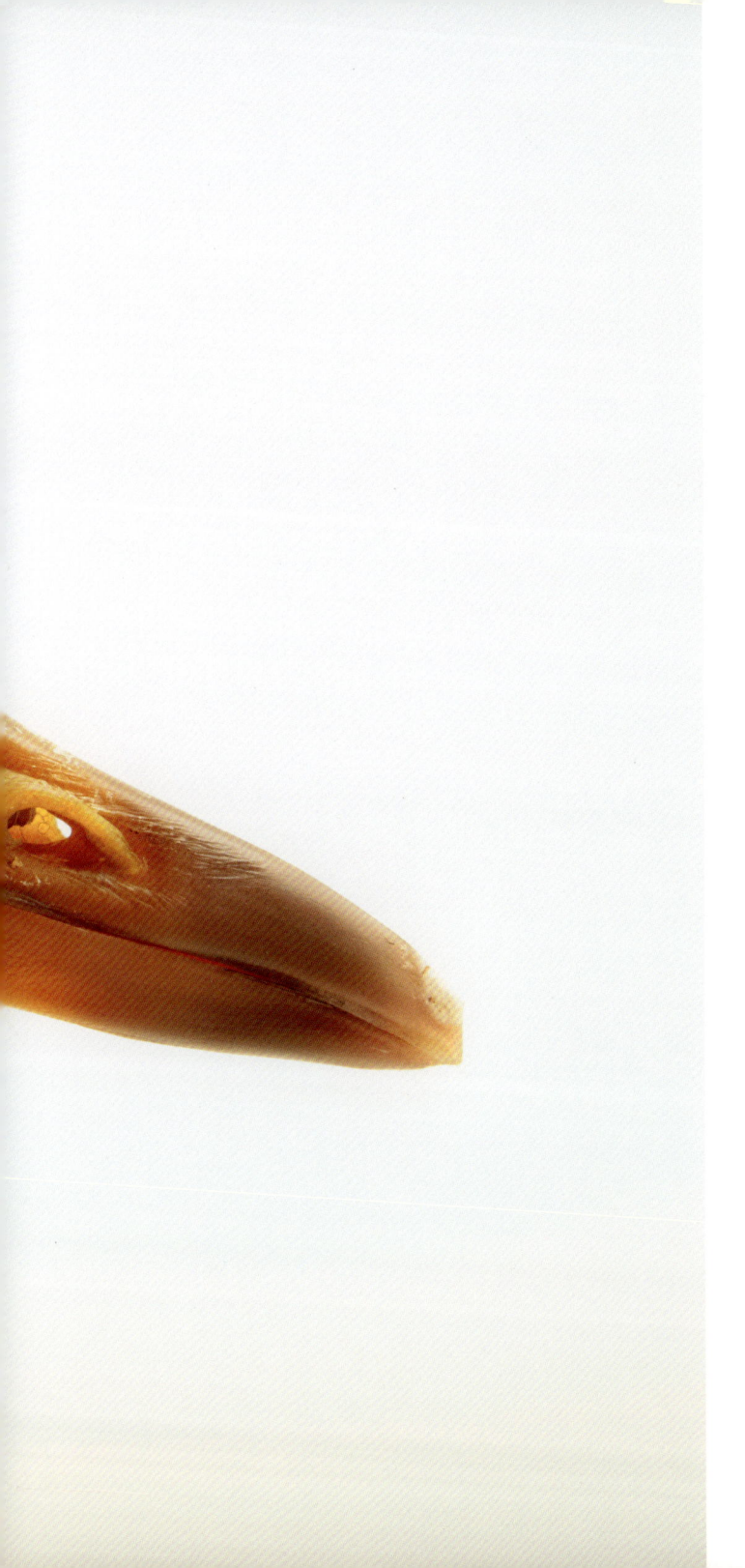

Greater sandhill crane

Antigone canadensis tabida (LC)

Sandhill crane chicks have very attentive parents. Males and females take turns incubating their eggs and brooding their chicks. Parents feed and protect their young until they are about 10 months old.

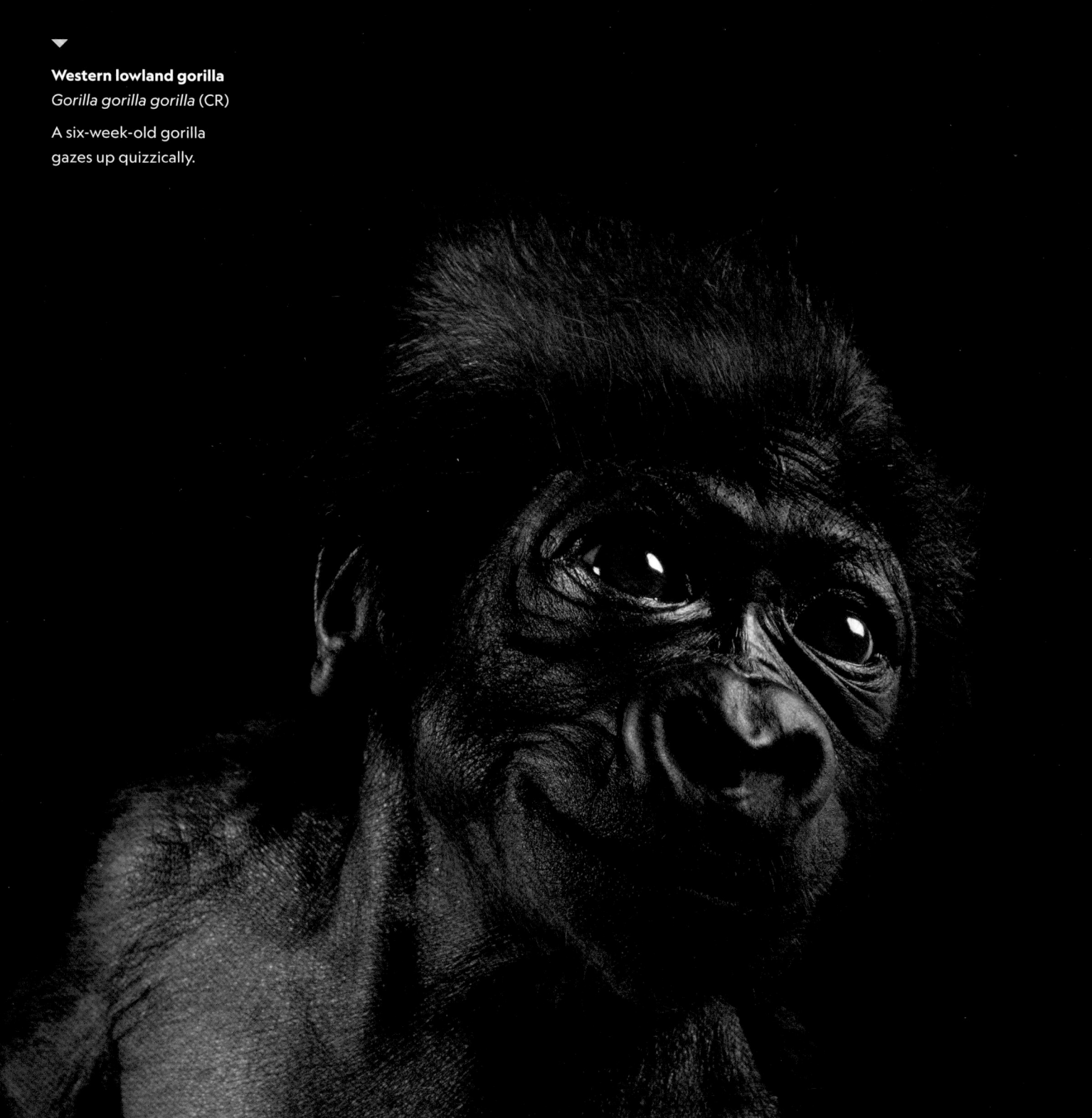

Western lowland gorilla
Gorilla gorilla gorilla (CR)

A six-week-old gorilla
gazes up quizzically.

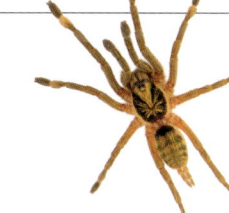
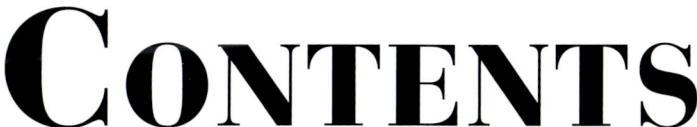

CONTENTS

INTRODUCTION 14

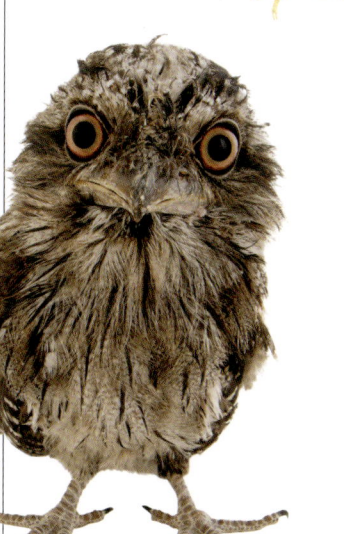

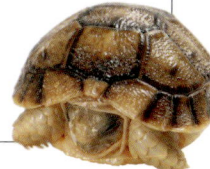

Clockwise from bottom left: **Humpback unicornfish** *Naso brachycentron* (LC); **tawny frogmouth** *Podargus strigoides* (LC); **linear cobalt crayfish** *Cambarus gentryi* (LC); **Humboldt penguin** *Spheniscus humboldti* (VU); **Trinidad olive tarantula** *Neoholothele incei* (NE); **red kangaroo** *Osphranter rufus* (LC); **Standing's day gecko** *Phelsuma standingi* (VU); **Kleinmann's tortoise** *Testudo kleinmanni* (CR)

INTRODUCTION

"In these photographs, time has simply stopped. That baby will always be exactly the same age and size, playful, full of curiosity and adventure."

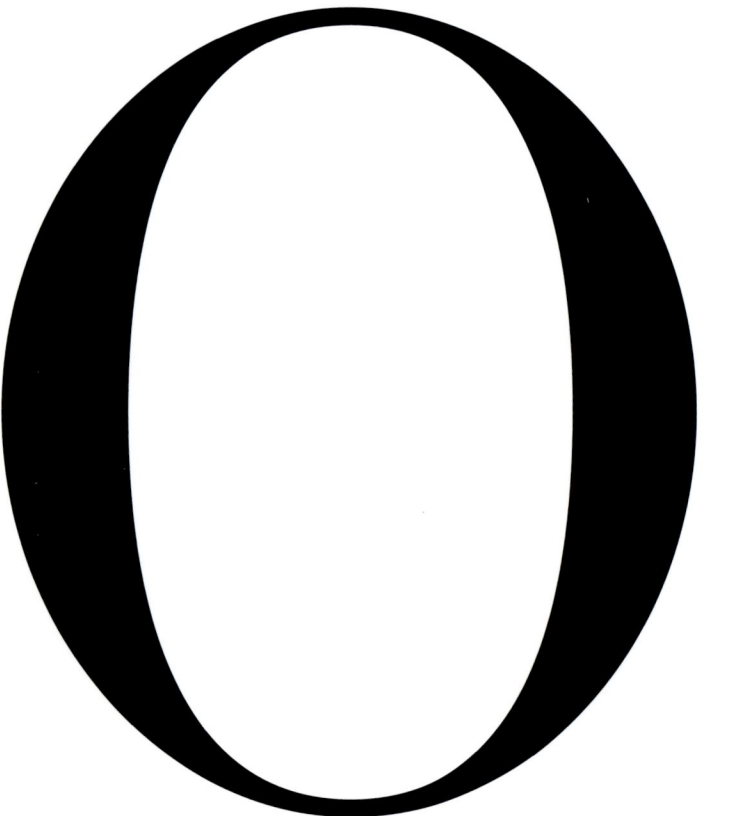

OF ALL THE IMAGES I've taken for the Photo Ark over the past 18 years, people often ask me, "What's your favorite?" I smile and say, "The next one." I've got to care about all creatures great and small, after all, and can't afford to discriminate.

But secretly, I think immediately about one image in the back of my mind. It's odd, because the photo is as simple as can be, but to me it stands above the rest.

It's the baby Malayan tapir at the Minnesota Zoo. (You can see her on pages 94–95.)

I often think about the day I met her. I was at the zoo to photograph other species, but as I wound down for the day, a couple of the zookeepers insisted I get a photo of their new baby, just six days old. I told them I already had an adult *Tapirus indicus,* shot at

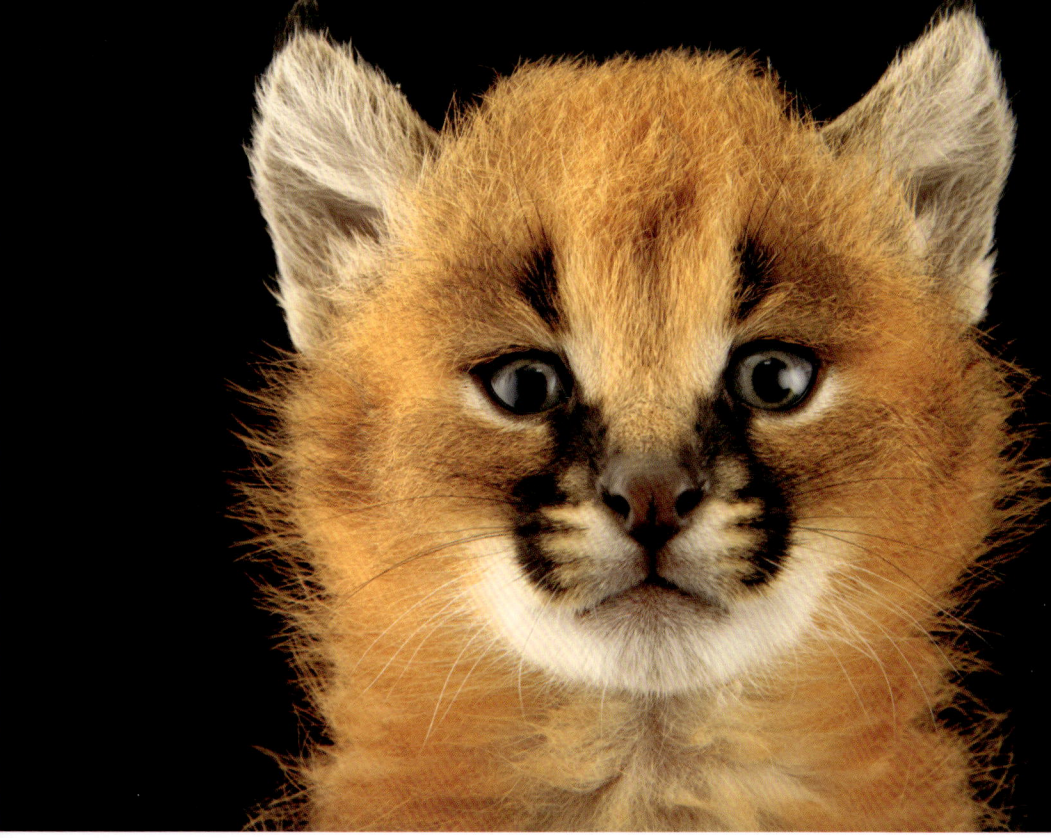

Caracal

Caracal caracal (LC)

This four-week-old kitten has a young caracal's characteristic blue eyes and short ear tufts. Watchful caracal moms look out for their kits until they're about six months old.

Omaha's Henry Doorly Zoo many years earlier, and that I'd never finish the Photo Ark if I documented all the baby animals.

Still, the keepers pressed on, so we made our way across the zoo and into a maternity stall in the tapir holding area. Everyone had huge smiles as they watched me round the corner, and for good reason.

There, in a simple room filled with straw, stood mother and child, as if they were expecting me. The mom, named Bertie, had been eating a bit, while the baby, named Amirah, stood there looking at me. And my heart just melted.

The young one was tiny, not much bigger than a loaf of bread. Not only was she cute, but she was also unlike anything I'd ever seen. Though Bertie was half white and half black, just as adult Malayan tapirs are famous for, her baby was black as night and covered with white spots and stripes. I couldn't believe my eyes.

This was no accident: Mother Nature designed these markings for good reason. In the wild, when lying down on a forest floor dappled with spots of sunlight, this baby would literally disappear. And because baby tapirs spend all day waiting for Mom to return from feeding, this camouflage is perfect for hiding from predators. The baby would look just like Mom in a few months, once large enough to fight (and flee) on her own.

The shoot took only a few minutes, then I headed out to my car. I kept right on smiling, though, thinking about how lucky I was, both that the keepers urged me to take pictures and that I'd witnessed something few ever get to see. That baby tapir has turned out to be one of the Photo Ark's most popular animals over the years.

Beyond that moment in 2011, I've thought ever since of how fortunate I've been to be the guy who's building the Photo Ark. I've literally had a front-row seat to the wonders of nature, young and old, big and small. Every day, I'm thrilled and amazed by who stares into my camera lens, especially when they're newly minted.

I've also learned that cuteness has zero to do with the mechanics of survival in the wild. While the tapir exhibited one strategy for hiding, there are myriad ways to overcome the vulnerability of youth. As I see it, if you're a youngster in this dog-eat-dog world, you have three main methods to make it: You can blend in, take off, or let your parents defend you.

Hiding makes a lot of sense for many animals. Shorebirds often scrape a simple depression in gravel or sand, then lay eggs that blend right in. Once hatched, their chicks are nearly as speckled and are camouflaged too. You literally can have a nest between your feet and miss it. Aquatic invertebrates can vanish under rocks on the floor of a streambed. Young geckos blend perfectly into the moss on a tree.

Running off is another option. Ungulates like wildebeests and zebras go from newborn to walking in minutes. Precocial birds from ostriches to prairie chickens hatch, dry, and are ready to follow Mom soon after. She'll teach them what to eat, how to shelter from storms, and how to escape from the bad guys.

To me, baby mussels are the most interesting of all. Mom coaxes a stream fish to come close by making a lure from her own flesh that looks tasty, like a worm or even a small fish. But that lure is a Trojan horse, filled with her tiny larvae. If a fish takes the bait, thousands of microscopic baby mussels explode out. Though only a handful are lucky enough to attach to the fish, those that do grab on will fall off their host weeks later to start a new life somewhere else. Mussels can't really move on their own, so this is their ingenious way to move the species upstream.

Then there's the way we humans and many mammals do it: long-term parental care. Like the baby tapir, newborn tigers stay put to nurse from Mom. Adults bring the youngsters solid food until they're old enough to hunt on their own. Monkeys, opossums, and bats literally carry their young around, providing safety and teaching moments along the way. Orca, dolphin, elephant, and gorilla offspring stay with their parents for years, just like we do.

All these strategies may seem obvious, but every one of them has been honed over time, handed down through thousands of generations. If it works, it stays. If not, it's gone. Simple as that.

And finally, one more baby story, from Omaha's Henry Doorly Zoo and Aquarium.

I'd long wanted to get a shot of an African savanna elephant against a white background, but it took a baby to make it happen.

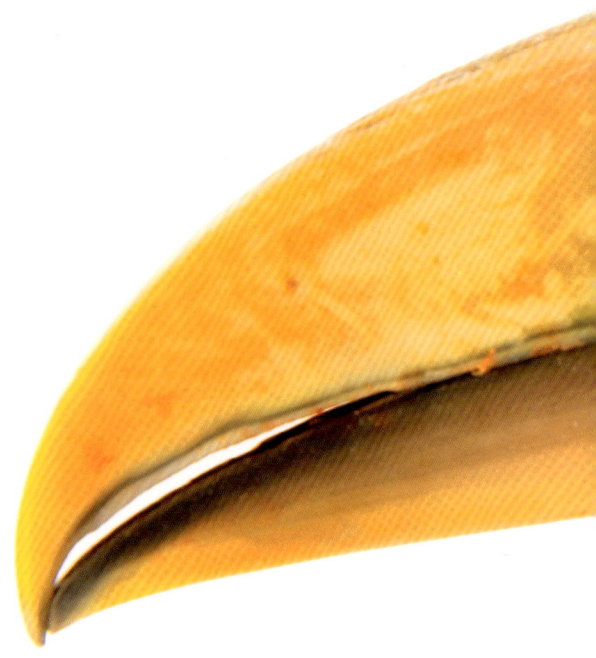

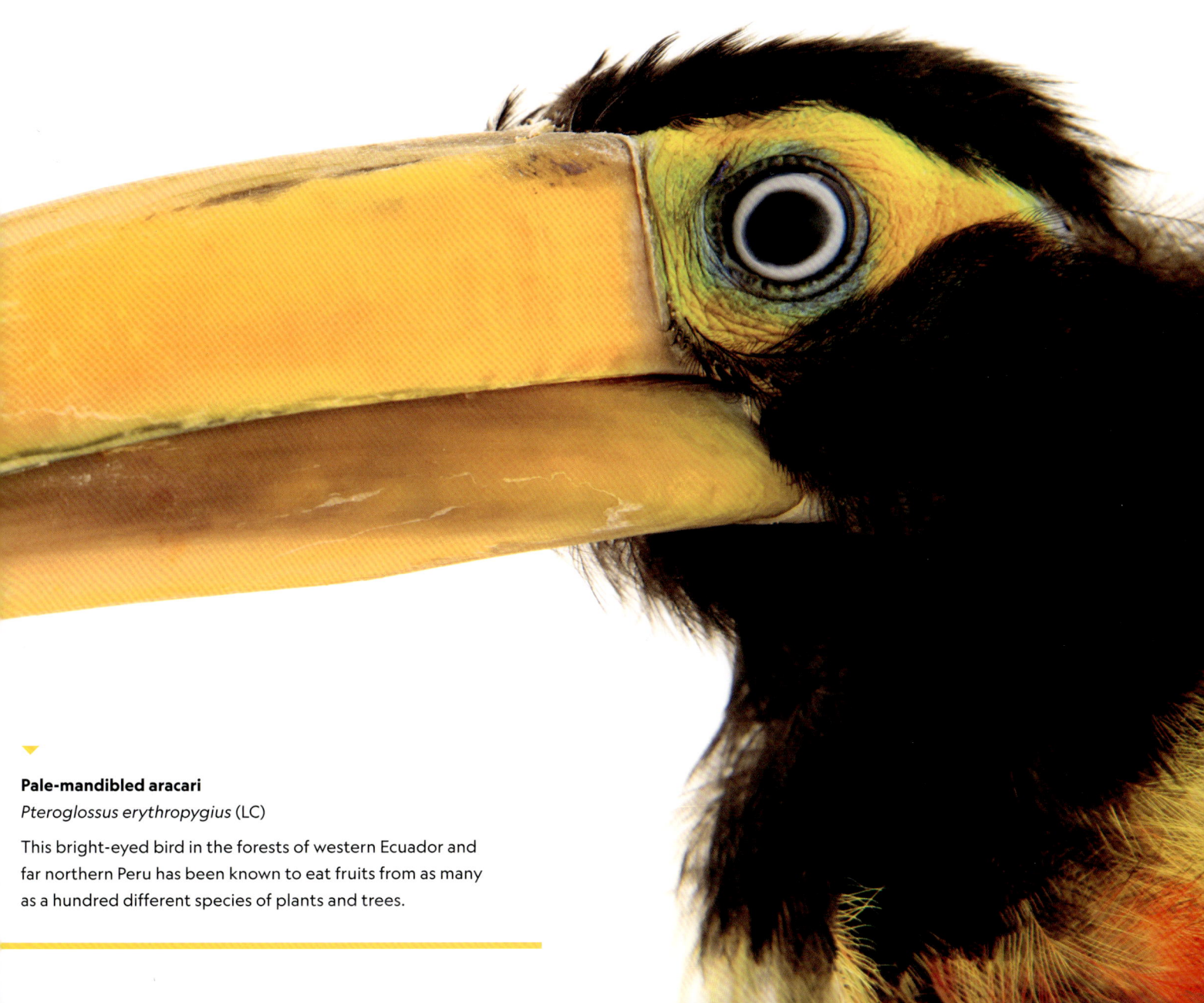

Pale-mandibled aracari

Pteroglossus erythropygius (LC)

This bright-eyed bird in the forests of western Ecuador and far northern Peru has been known to eat fruits from as many as a hundred different species of plants and trees.

My friends at the zoo told me again and again how "stinkin' cute" their brand-new baby elephant was. Being the busy guy that I am (and knowing how much work it would be to create a white space big enough to fit an elephant), I dodged the shoot as long as I could. But when I finally did drop by, I stepped into the elephant barn and there she was. Just adorable.

Once the zoo agreed, we set out to paint a space white in the back of the house. I bought some special light stands that went up some 20 feet (6 m) to project light downward, hoping to bathe the subjects in a heavenly glow.

The next day, keepers used food treats to get the mom (named Jayei) to come up to the white background and even center herself on it. The baby, Amandra, was never far away, orbiting the way an electron circles a nucleus. Mom stayed put thanks to the keeper's rewards. Eventually, the baby passed between Mom's legs a few times, and we had our shots (one of which you can see on pages 154–155).

The shoot was over quickly, the paint was power-washed off, and all was back to normal. We moved on.

In the resulting photographs, time has simply stopped. That baby will always be exactly the same age and size, playful, full of curiosity and adventure. And yes, just as cute as she can be.

I suppose this business of freezing time is what drew me to become a photographer in the first place. In a whirling world, these little moments, each 1/250th of a second, present life in a way we can rarely see for ourselves.

But in these pages, we have all the time we need to realize how amazing every life-form is, from insects to the world's largest land mammals. I wonder what the future holds for the babies in this book. Will they tug at enough heartstrings to inspire us to improve the planet at long last? Will we humans realize what a gift this all is and start saving the wild places that elephants (along with every other creature) need to thrive and raise their young?

I sure hope so. After all, when we save other species, we're actually saving ourselves.

66

"People have to fall in love with these animals if we ever hope to save them, and there's no better way to do that than with baby pictures."

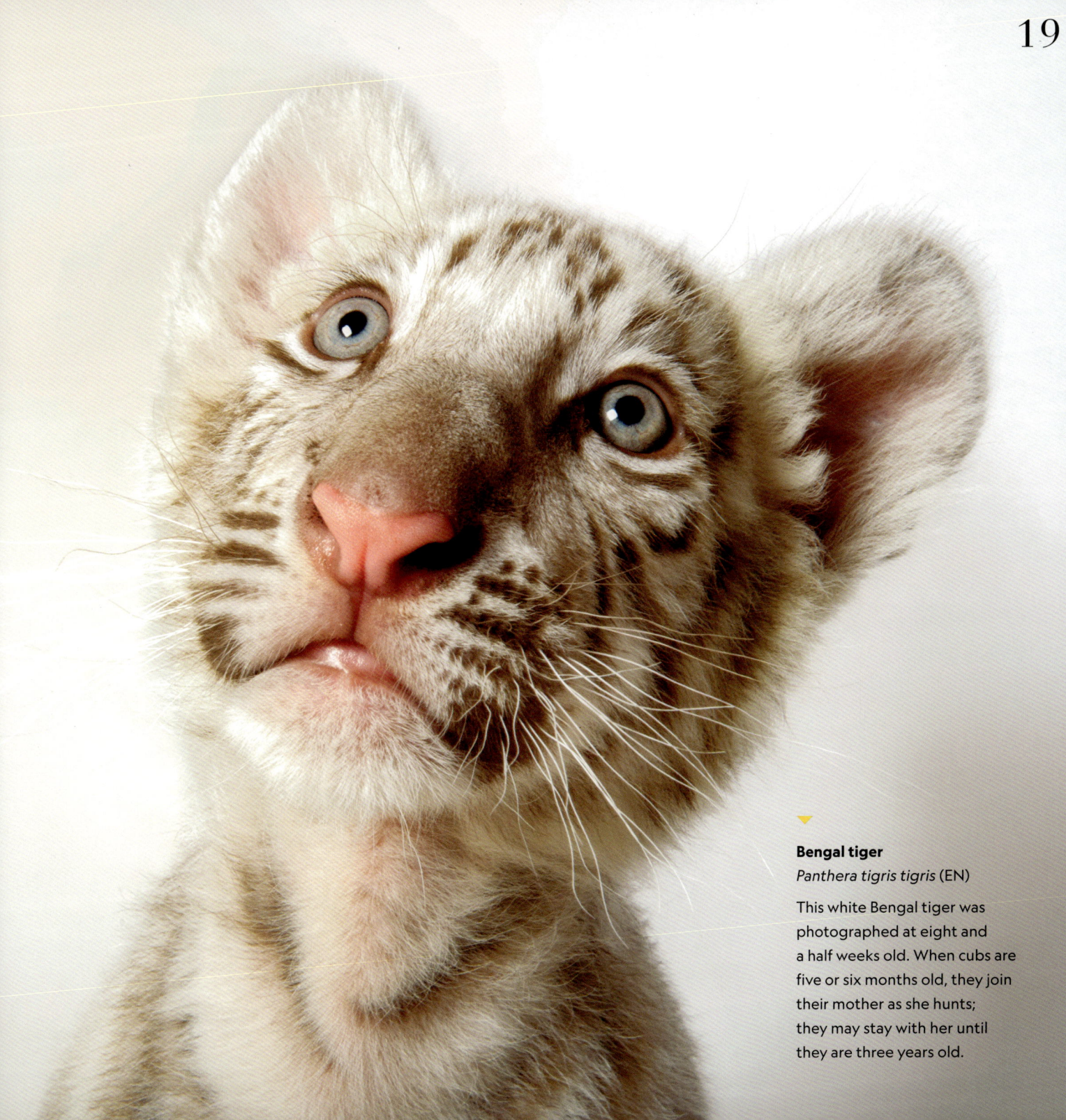

Bengal tiger
Panthera tigris tigris (EN)

This white Bengal tiger was
photographed at eight and
a half weeks old. When cubs are
five or six months old, they join
their mother as she hunts;
they may stay with her until
they are three years old.

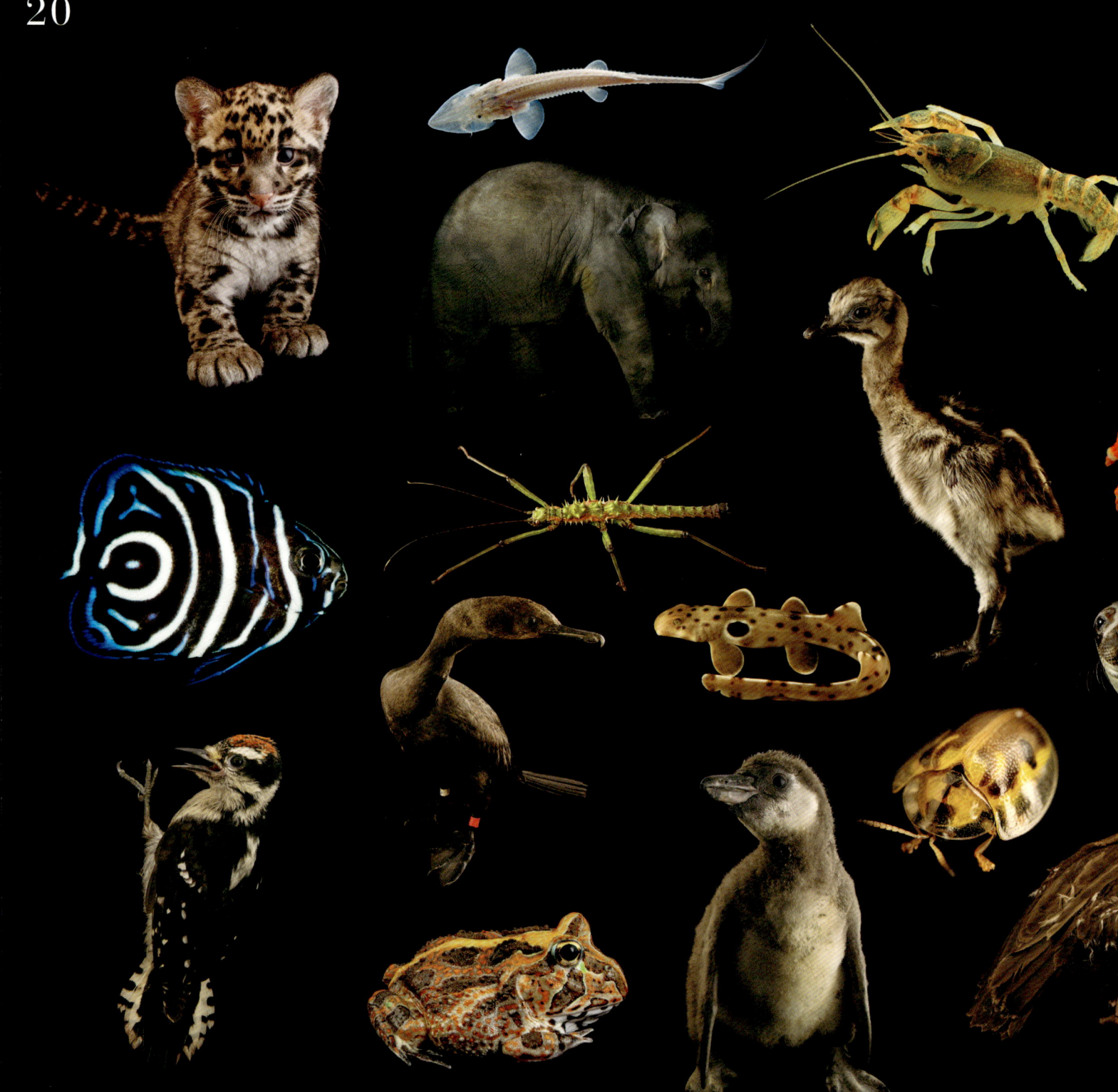

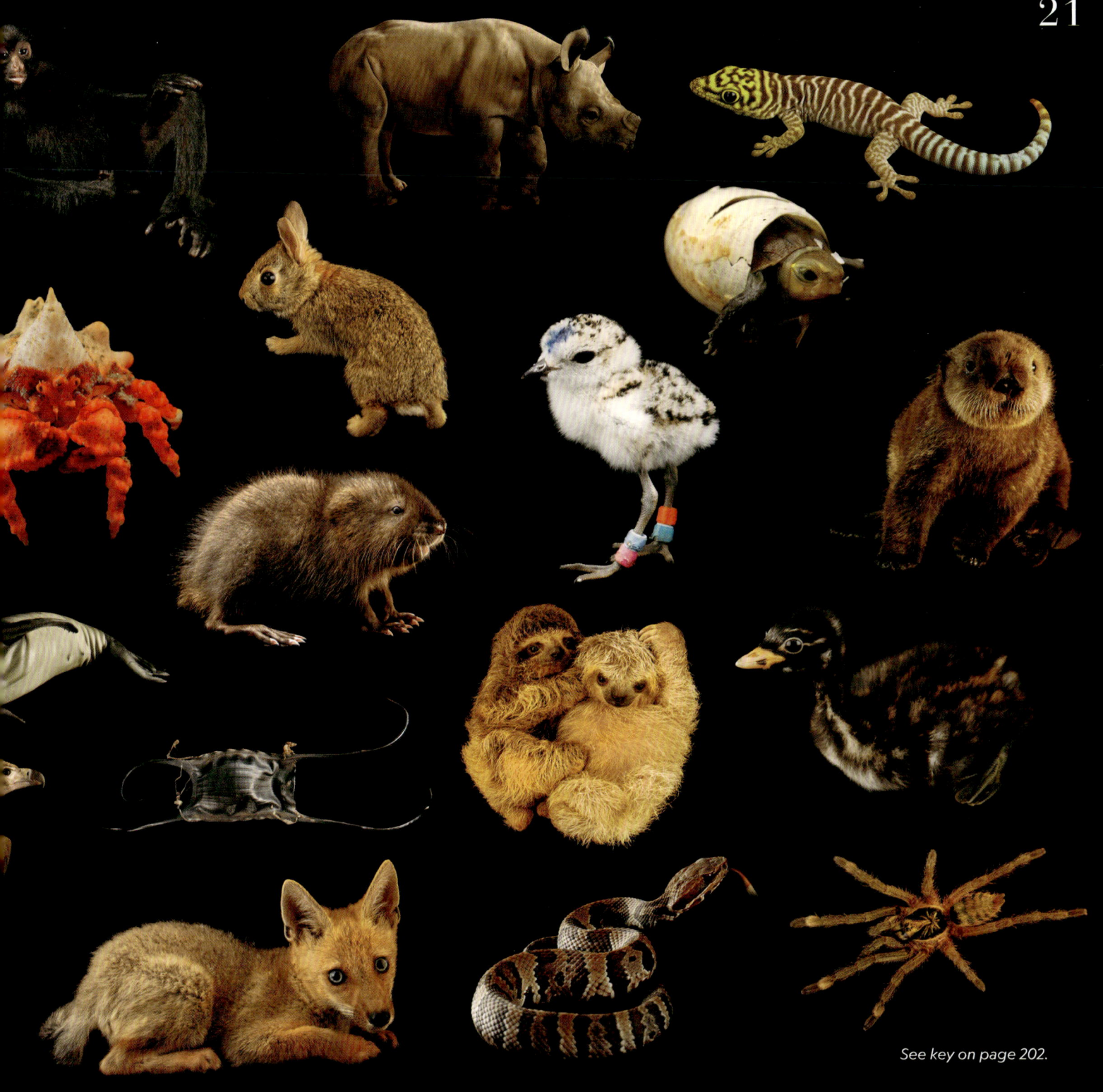

See key on page 202.

1 | BEGINNINGS

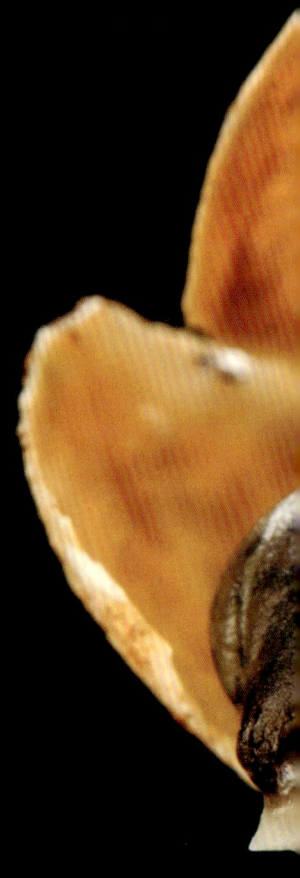

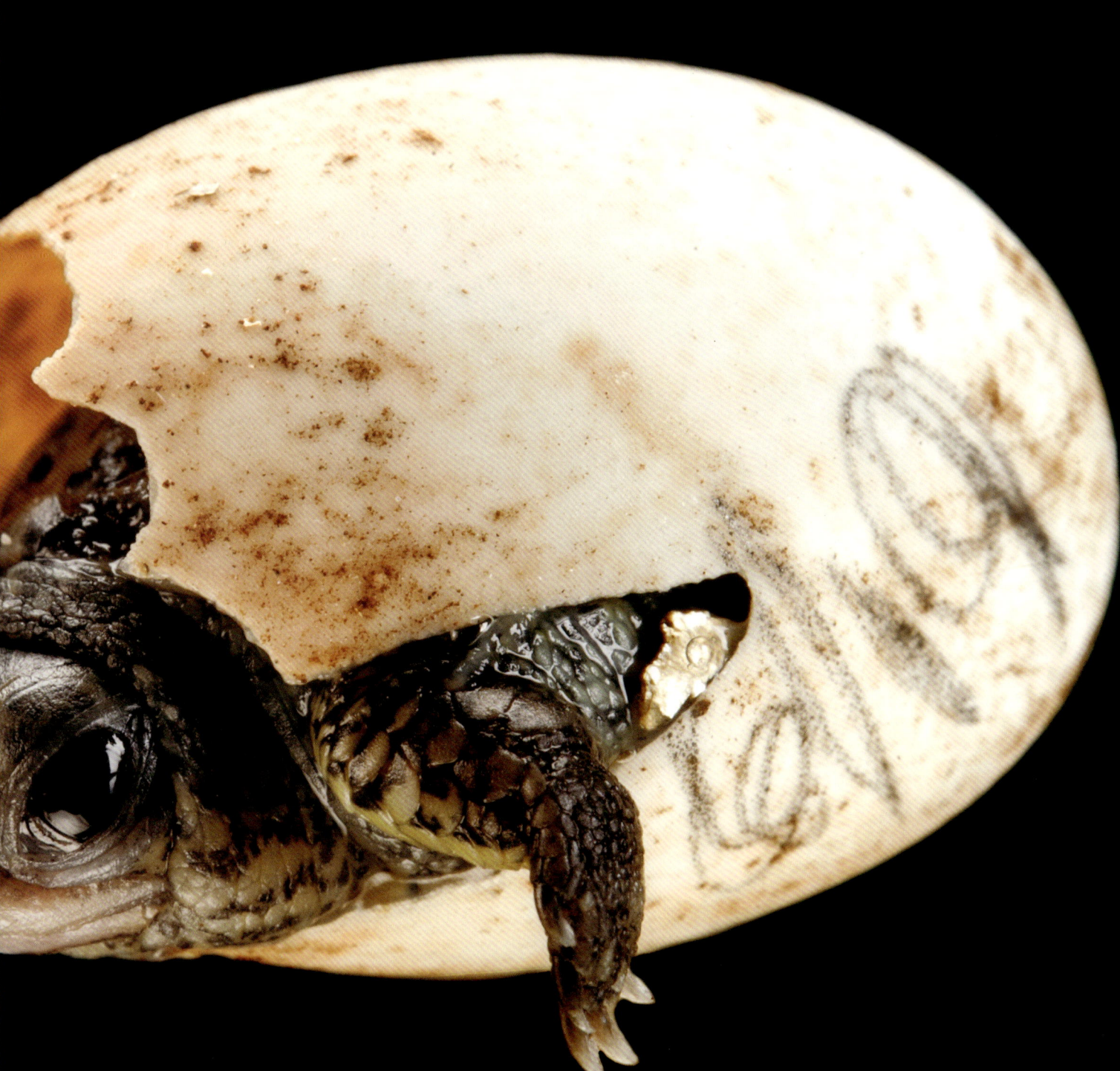

THE STARTING LINE

New life is a marvel to behold. Whether we ponder a bear cub born while its mother hibernates or tiny spiderlings ballooning into the breeze on gossamers of silk, beginnings are often tinged with mystery and magic.

B**EHIND THE WONDER LIES** a grand strategy. An animal's drive to perpetuate its species is utterly primal. By combining genes, animals fare better in the long game of natural selection. Fluffy chicks and wide-eyed infant monkeys, in other words, are charming evidence of nature hard at work.

Life begins differently across the animal kingdom. On one end of the spectrum are African savanna elephants, who reach sexual maturity at around 10 to 12 years old and have the longest pregnancy of any living mammal. For nearly two years, the mother carries her growing baby inside and, after giving birth to a calf, may not have another for several years. Other animals breed early and often. Eastern cottontail rabbits, for instance, can reproduce when they're only two or three months old, and one female can give birth to as many as seven litters a year. Egg layers can also be prolific: They lay, release, or deposit eggs—a handful, a few hundred, tens of thousands at a time—and with luck, enough will hatch and survive to keep the species going.

Eggs may be common, but they are one of nature's most miraculous inventions. Built to nurture and protect the growing life within, eggs are everywhere, in all shapes and sizes: in leafy treetops, in backyard streams, stuck to the underside of a tropical leaf, and even in poop. Each of the world's 10,000-plus species of birds lays a different kind of egg; our planet's one million species of insects produce eggs in staggering numbers. Yellow-headed geckos deposit their eggs in leaf litter; some tree frogs, in a nest of foam; Colorado river toads, in a tube of jelly. Eggs can be hard-shell, soft and jiggly, or, like the mermaid's purses in which catsharks and skates develop, a special shape and texture all their own.

Although every species has its own unique way of reproducing, their life cycles often depend on one another. Like Atlantic horseshoe crabs and migrating shorebirds: Thousands of horseshoe crabs gather each spring during high tides of the full and new moons in Delaware Bay, an estuary of the mid-Atlantic. There, they mate and lay their eggs. A single female can lay up to 20,000 eggs in one night. This phenomenon happens at just the right time and in just the right place for migrating shorebirds like red knots—bound for Canada to breed and raise their own young—to pause and feast on the horseshoe crab eggs.

It's a boon for the birds, but the coordination checks out for the crabs too: They've been around for roughly 300 million years. The crabs, the shorebirds, the phase of the moon—all in sync to play their part in the ongoing circle of life.

▼

Monarch butterfly chrysalis *Danaus plexippus* (LC)
Previous pages: **Coahuilan box turtle** *Terrapene coahuila* (EN)

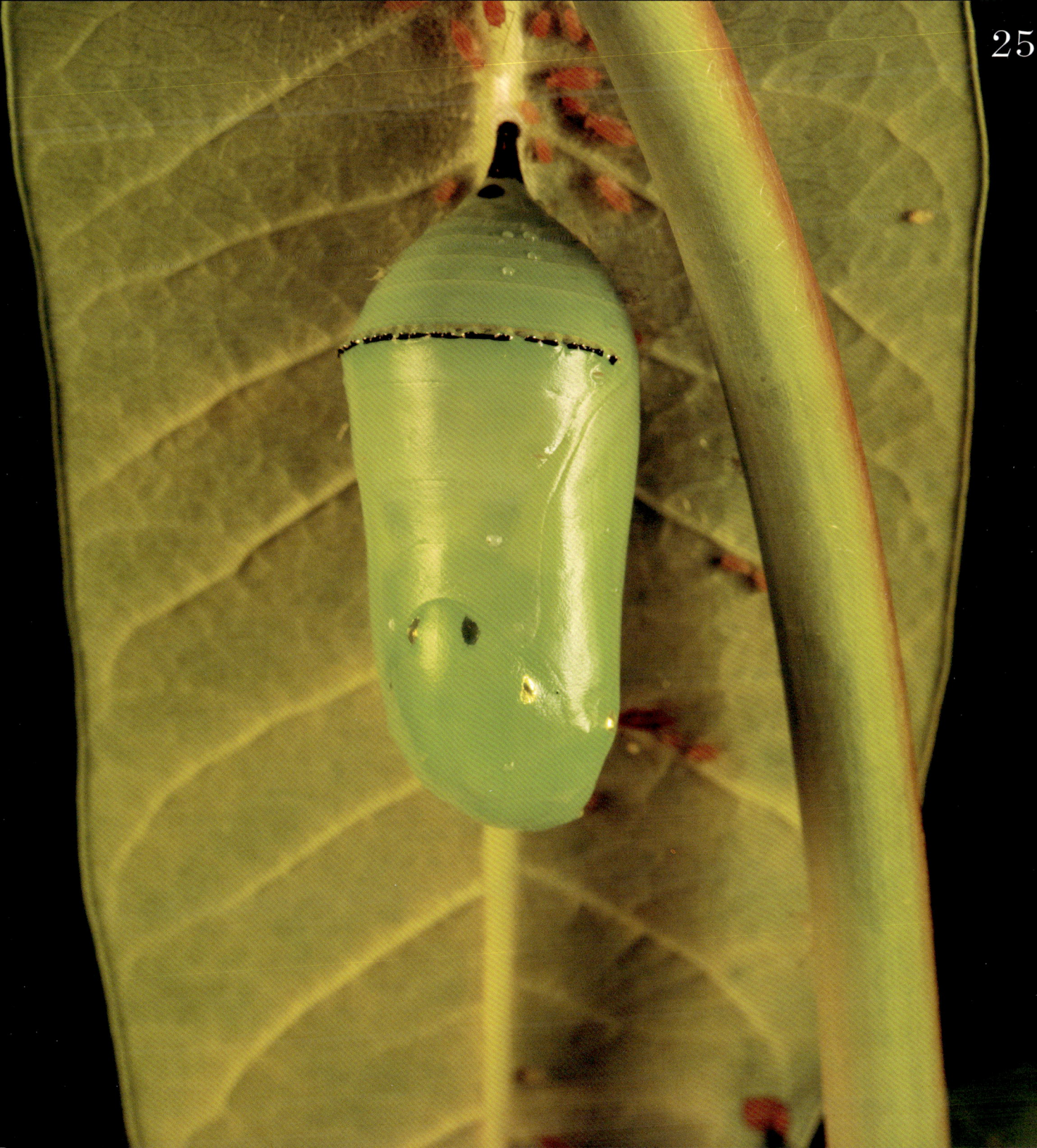

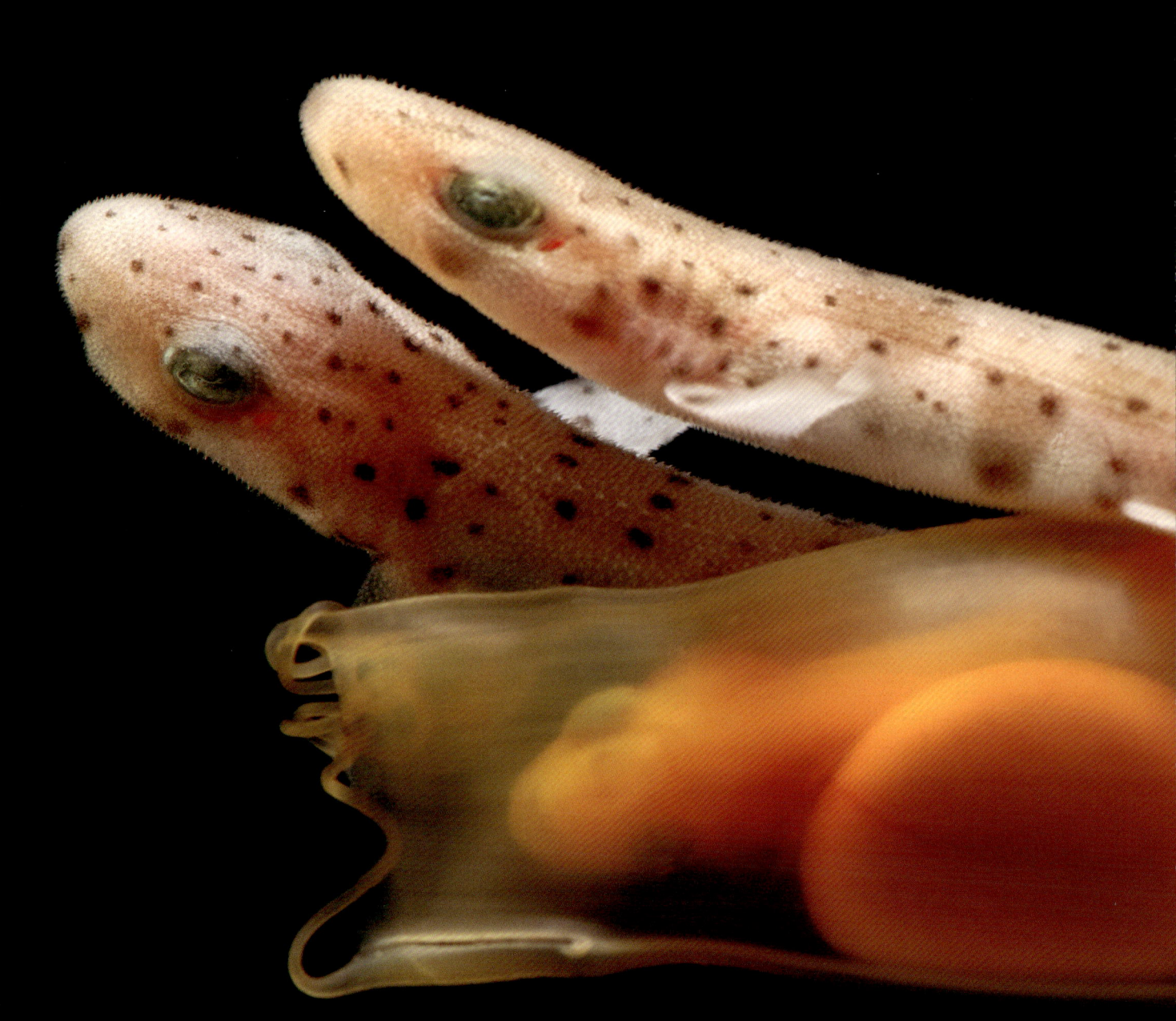

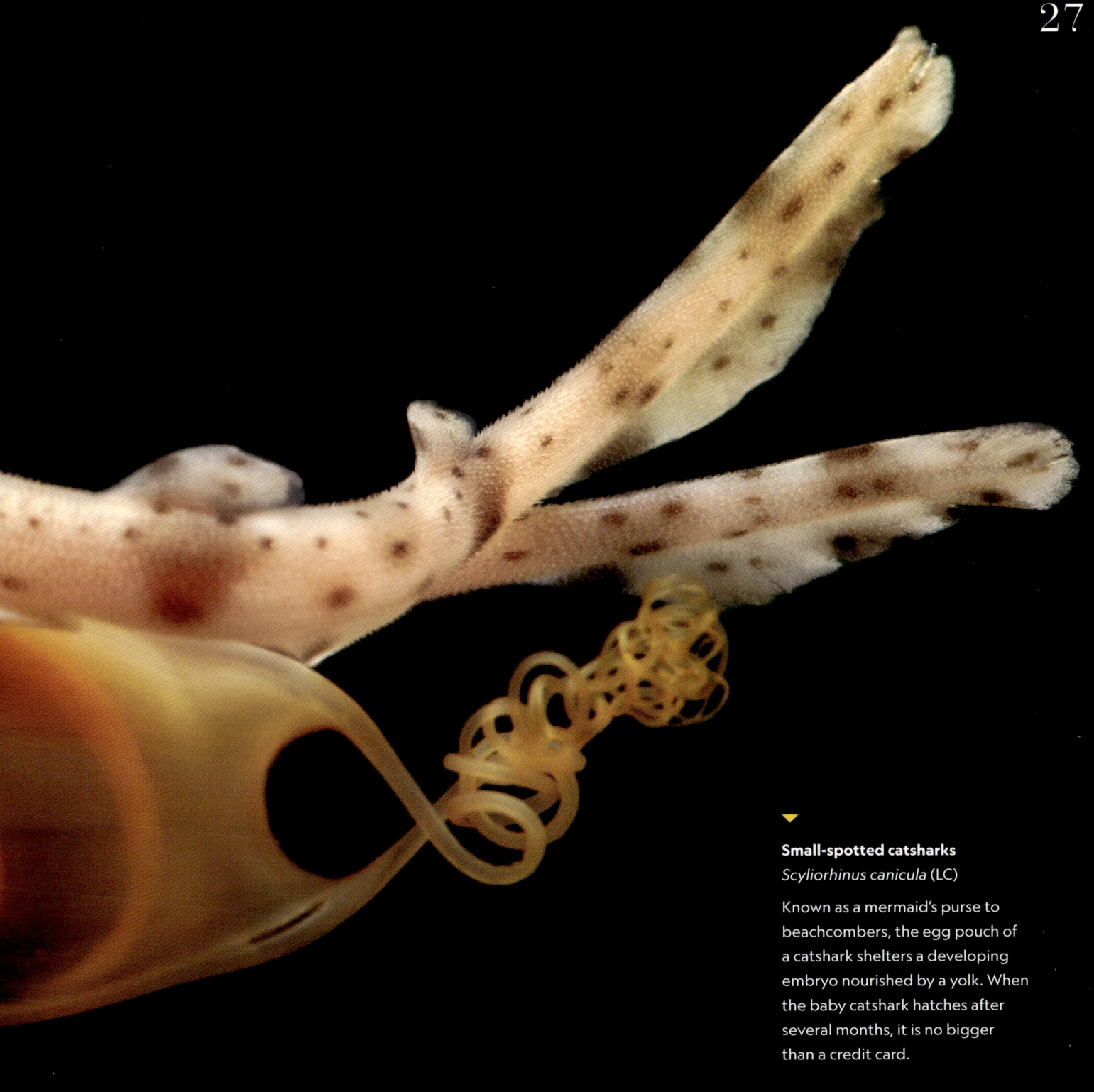

Small-spotted catsharks
Scyliorhinus canicula (LC)

Known as a mermaid's purse to beachcombers, the egg pouch of a catshark shelters a developing embryo nourished by a yolk. When the baby catshark hatches after several months, it is no bigger than a credit card.

ON TOP OF IT
Eggs Down Under

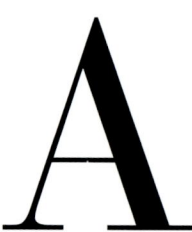

A CRAYFISH MOM is very watchful. She keeps her many eggs—numbering anywhere from 20 to 700—attached to her abdomen at all times to protect them from predators. She shields the eggs by tucking her tail and fanning it to aerate them. But it's a different story when the babies hatch. Although they still need their mother's protection for the first few weeks of life, a bit of biological magic needs to happen first. The mother crayfish must secrete a pheromone that will entice her to continue caring for her offspring. As soon as she stops secreting this pheromone, the babies have to dash: If they don't scurry away at once, she'll make an easy meal of them.

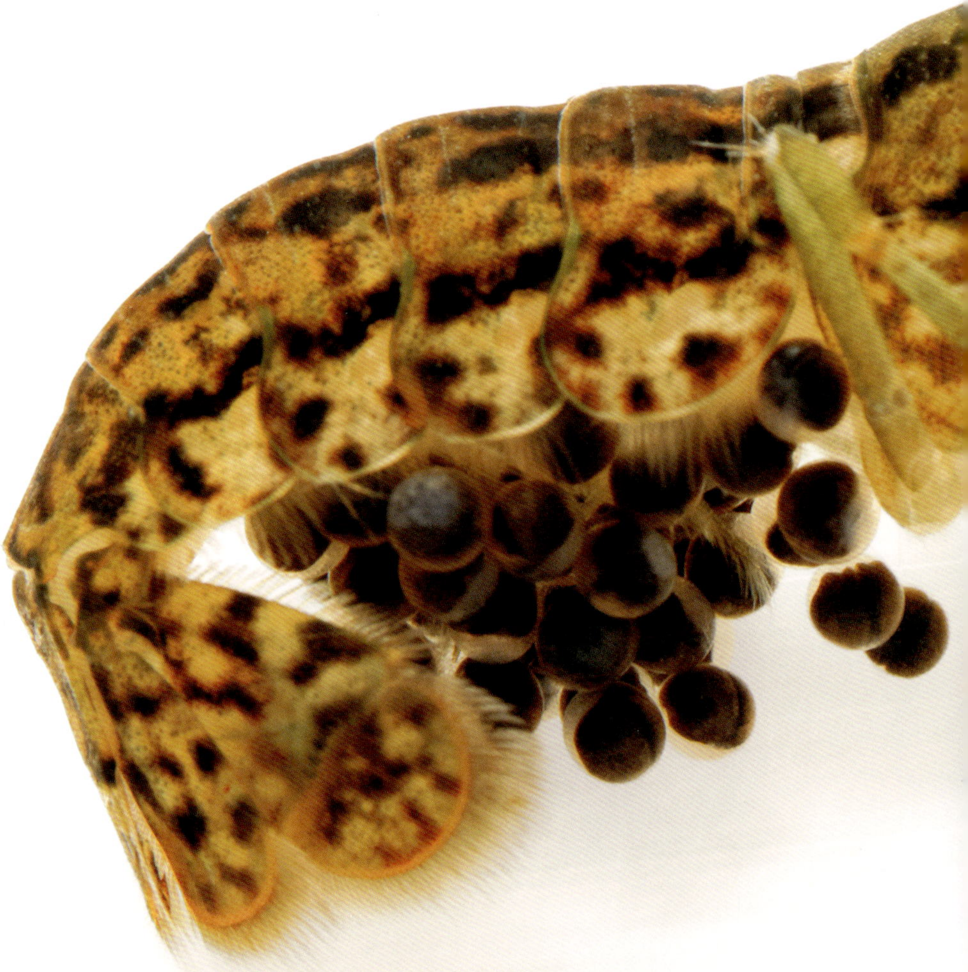

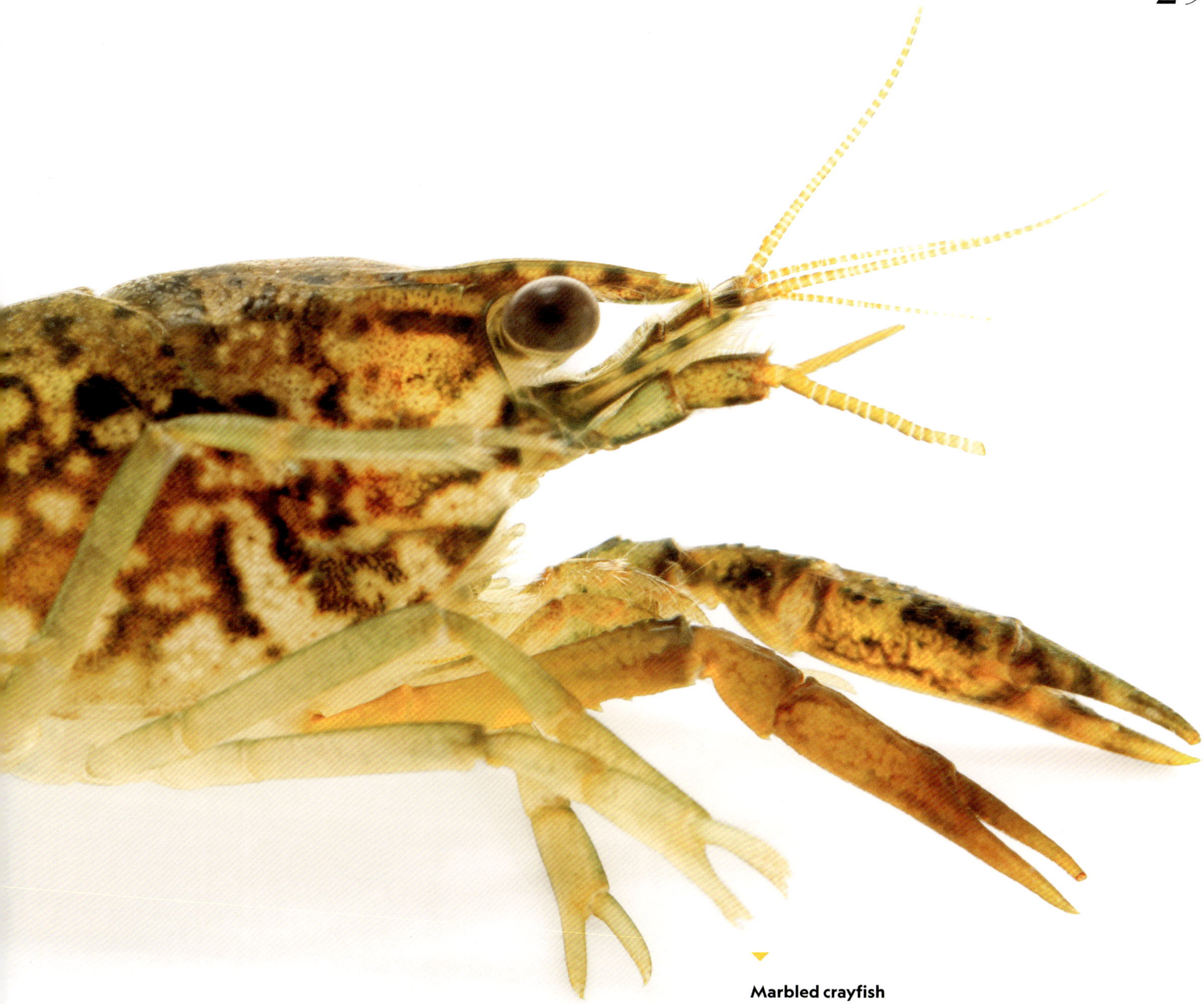

Marbled crayfish

Procambarus virginalis (NE)

More than half of the world's crayfish species live in North America. A female crayfish carrying eggs is sometimes called "in berry" because the eggs look like clusters of berries affixed to her abdomen.

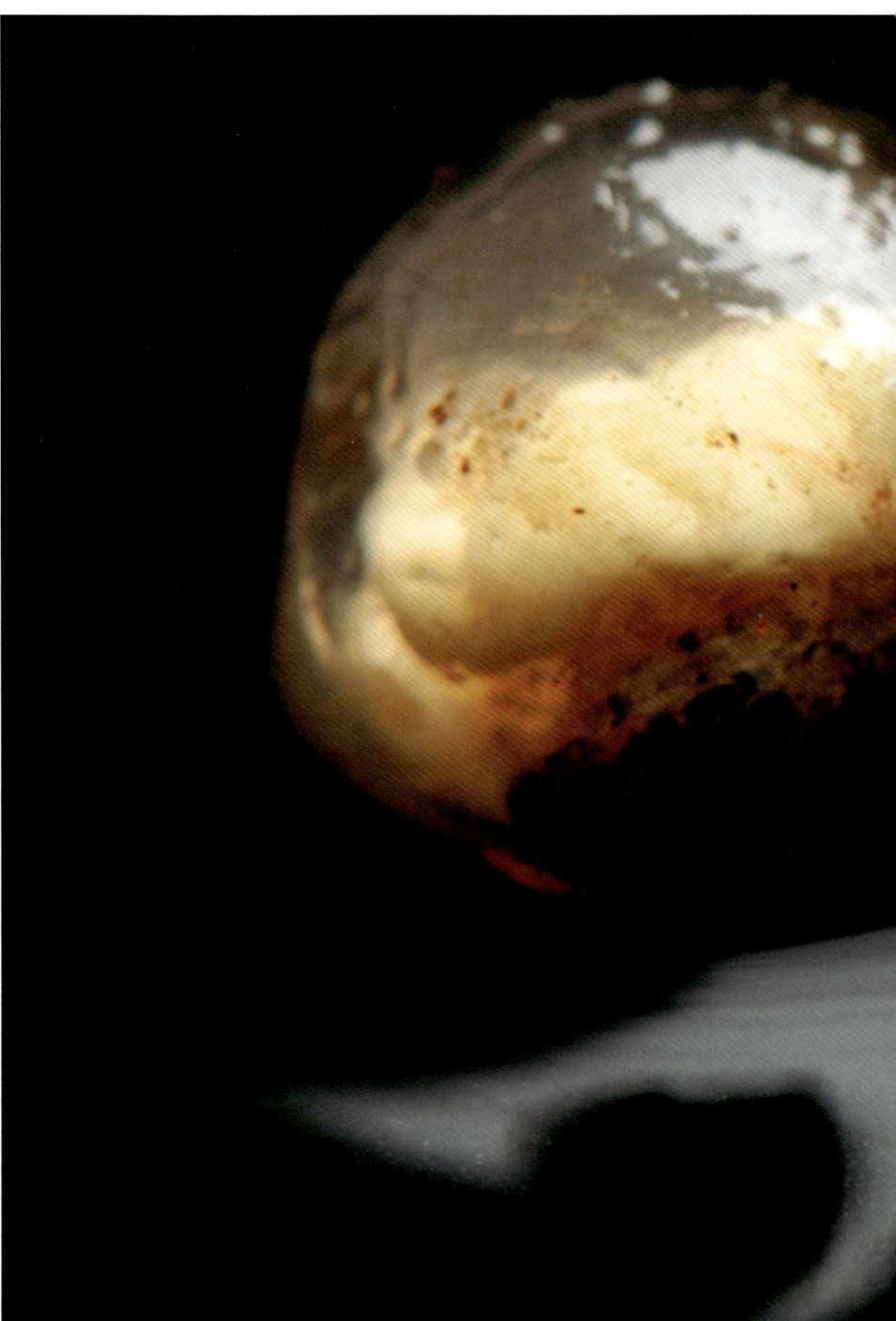

Solomon Island leaf frog eggs
Cornufer guentheri (LC)

The eggs of this species, native to the Pacific islands for which it is named, will hatch and produce fully developed frogs. It is one of only a few frog species that has no tadpole stage.

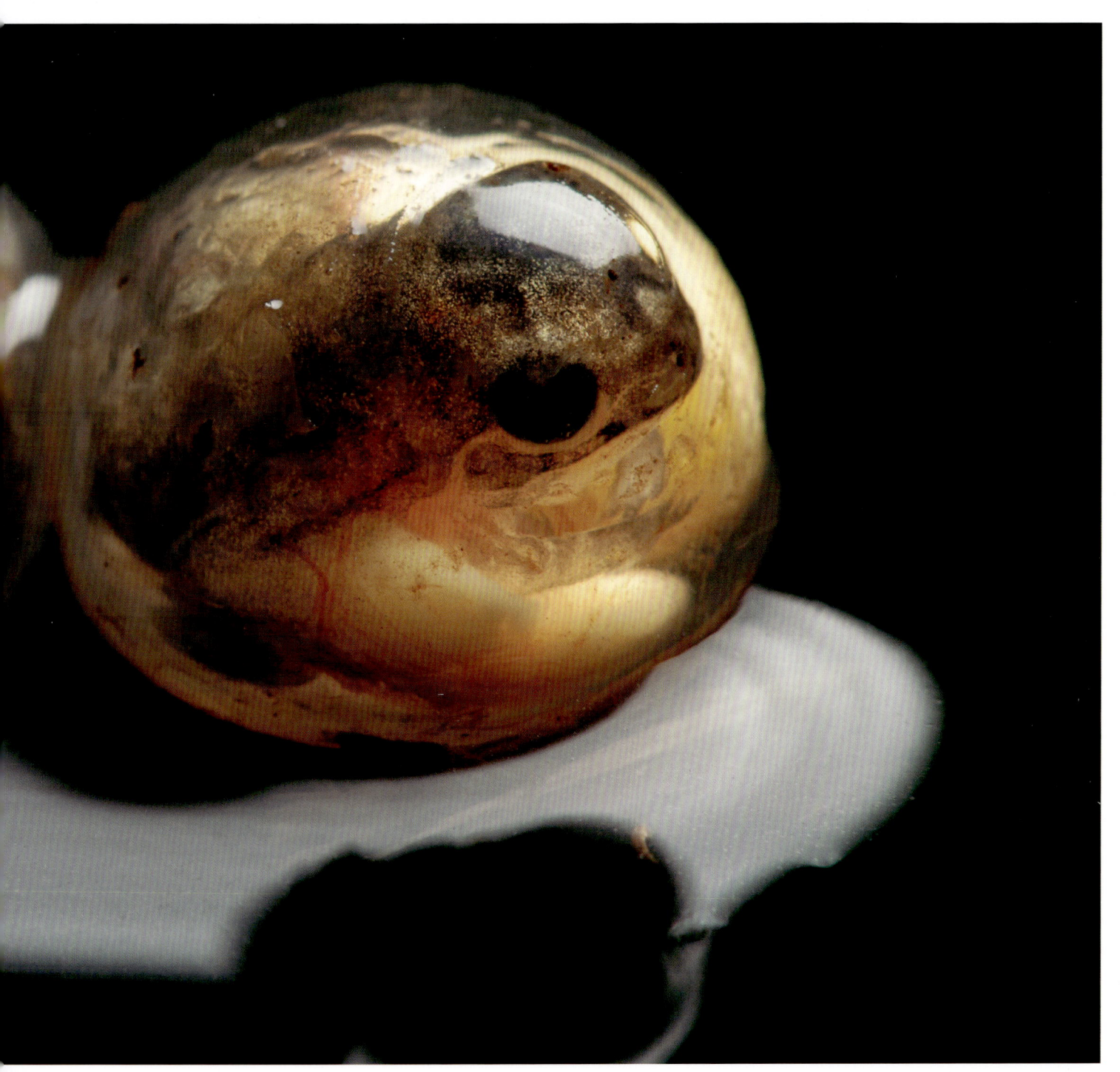

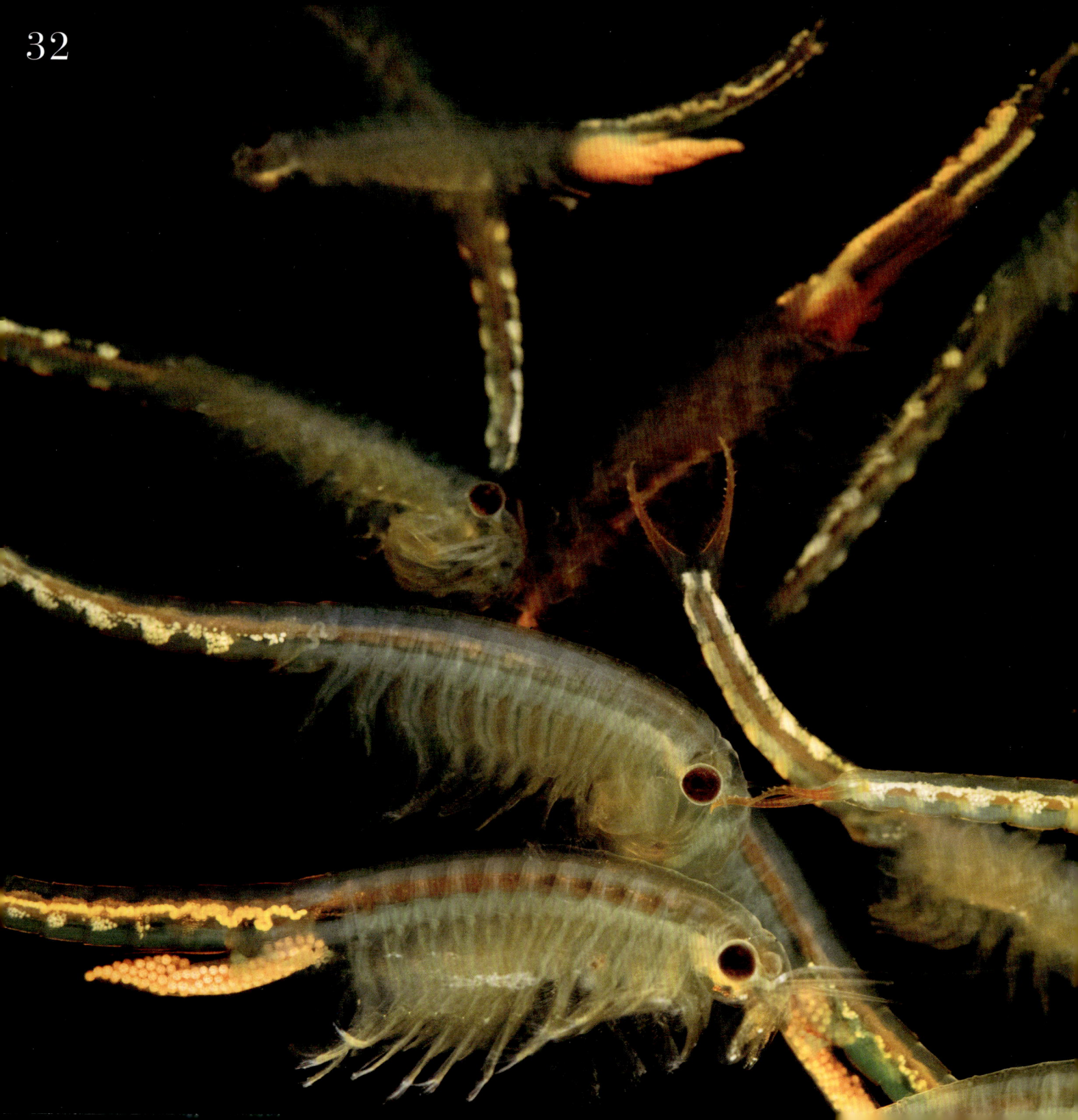

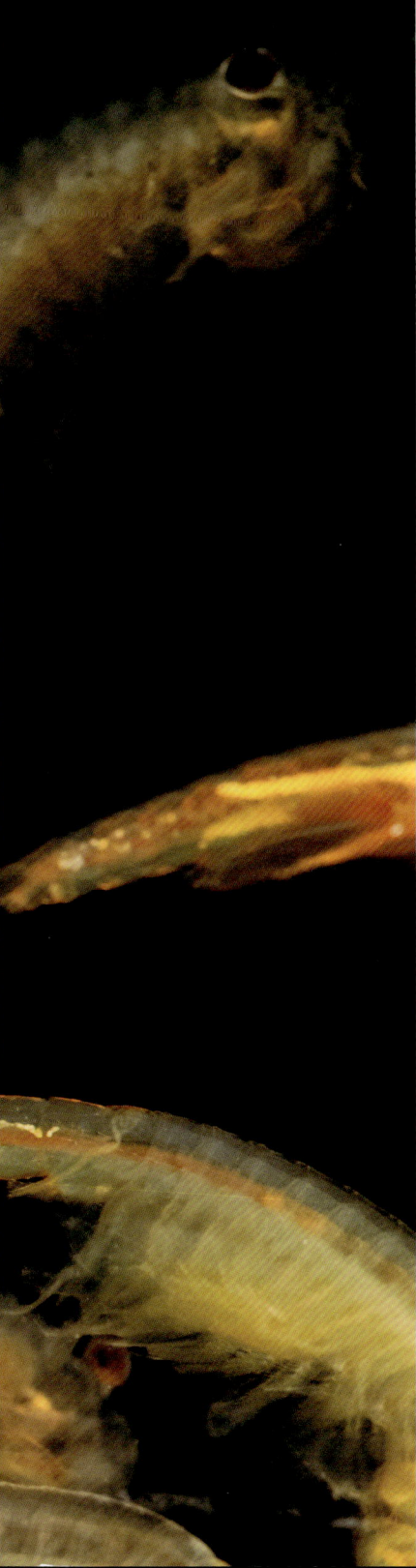

> "I certainly like to look at them all: baby birds, mammals, reptiles, insects, fish, and even these tiny shrimp. They are just as cute as can be, and so worth saving."

Fairy shrimp

Streptocephalus sealii (NE)

Fairy shrimp live in seasonal ponds and lay their eggs before warm weather comes and the pools dry up. The eggs (seen here as glowing golden bundles) stay dormant during the dry season, awaiting the flush of fresh water during the rainy season, before hatching.

American burying beetle larvae

Nicrophorus americanus (CR)

Larvae of this carrion beetle feed on carcasses their parents provide. The beetle once thrived across the United States before habitat loss and other threats caused its population to plummet.

DIRTY WORK

When an American burying beetle finds a dead animal—a mouse, a pheasant, or even a prairie dog—it joins with a mate and buries the critter, removing its fur or feathers. The pair then cover the carcass in bodily fluids to preserve it for their future brood of larvae. The female lays her eggs nearby, and when the larvae emerge they feed on the carrion until they mature and head out to find their own meals.

In studying this beetle species' unique life cycle and breeding behavior, scientists now find that human development and a general decline in biodiversity mean that fewer bird and rodent carcasses end up in prime burying beetle habitat, making it difficult for them to find the right conditions for breeding. Conservationists have reintroduced the insect to parts of its former range and use their knowledge of breeding behavior to shape conservation projects and maintain wild populations of the beetle.

The habits of the burying beetle may seem a bit dark. But a scavenger is also a recycler, and North America's largest carrion beetle is vital to the health of many ecosystems, returning nutrients to the soil and, by doing away with dead animals, limiting living creatures' exposure to disease.

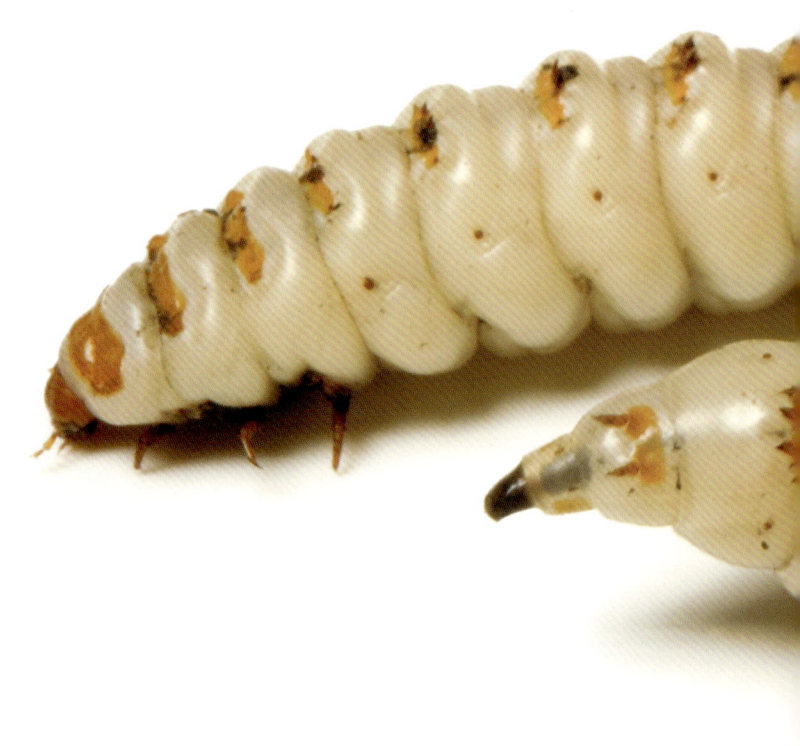

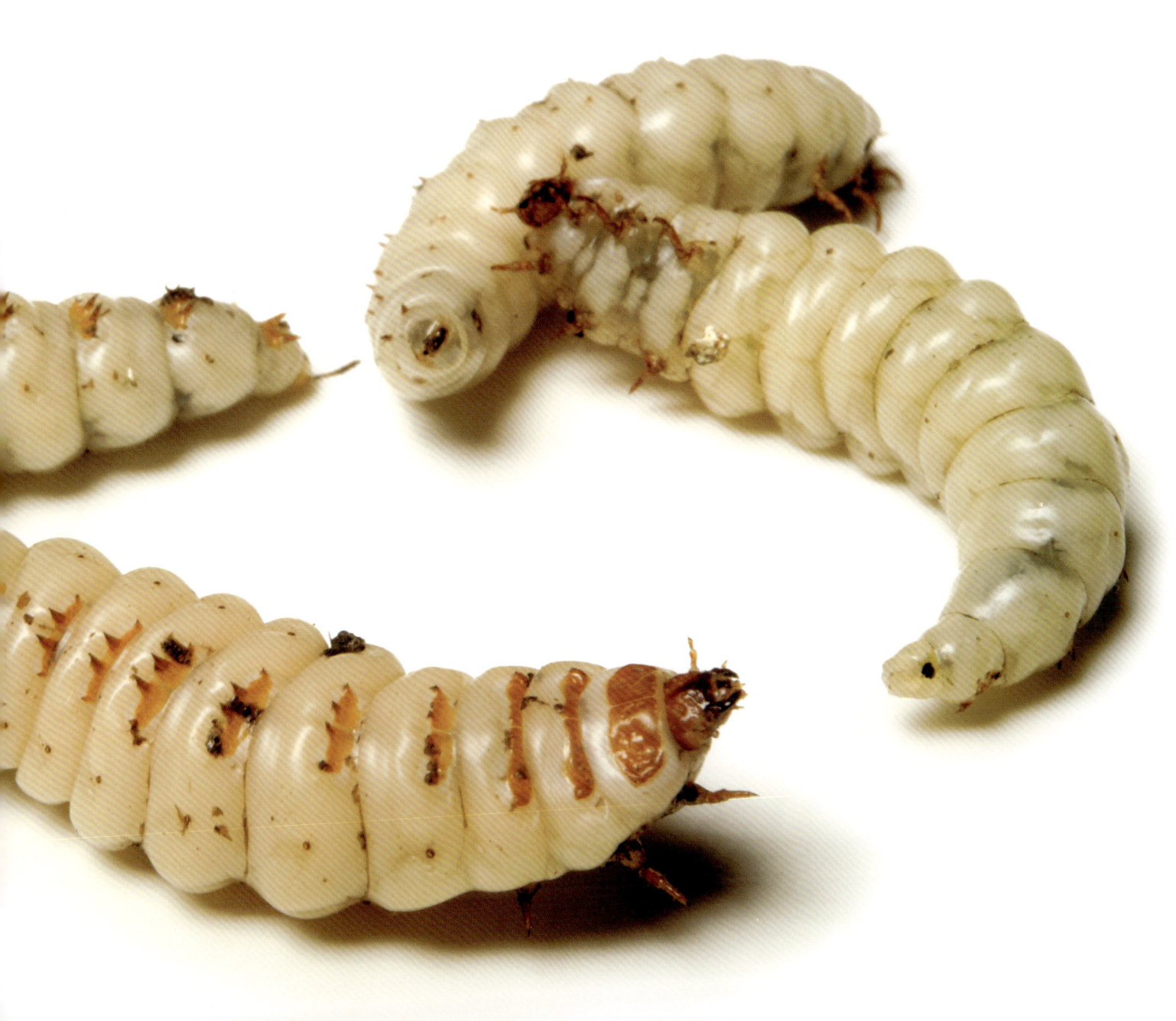

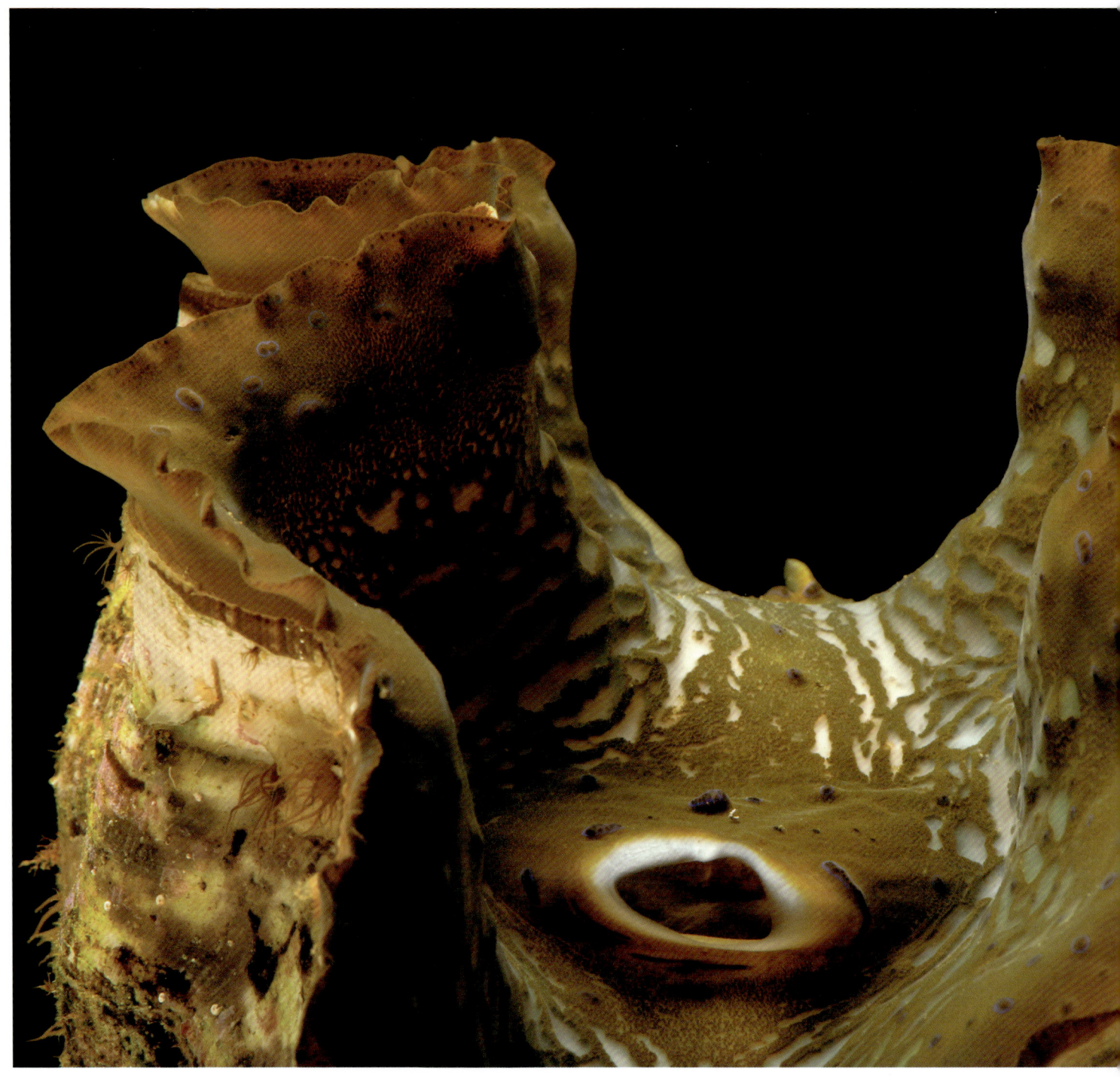

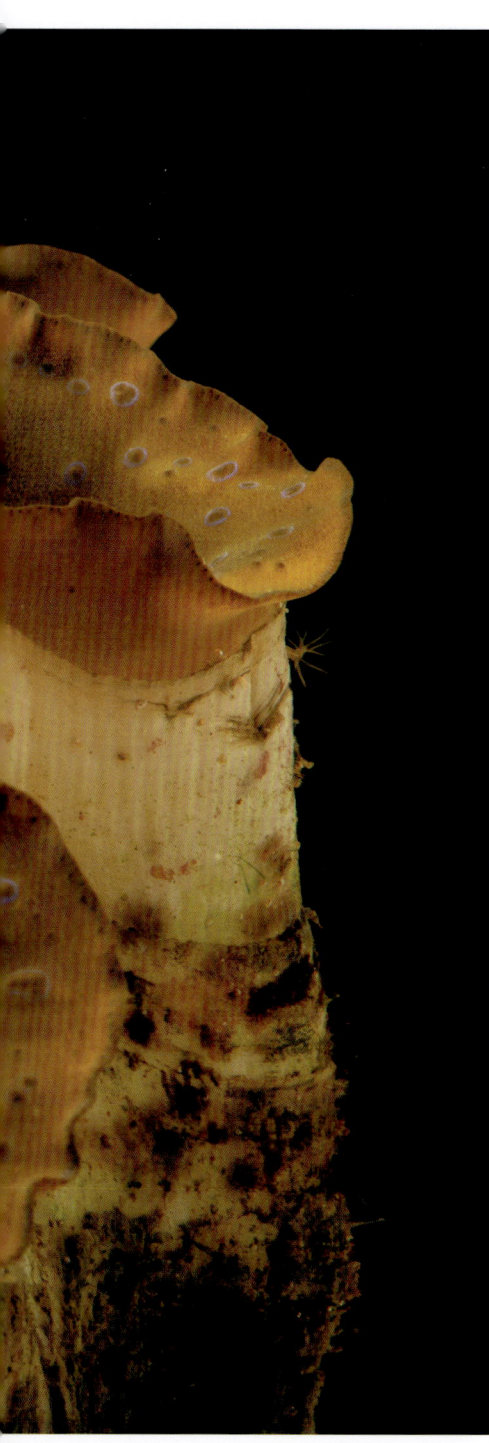

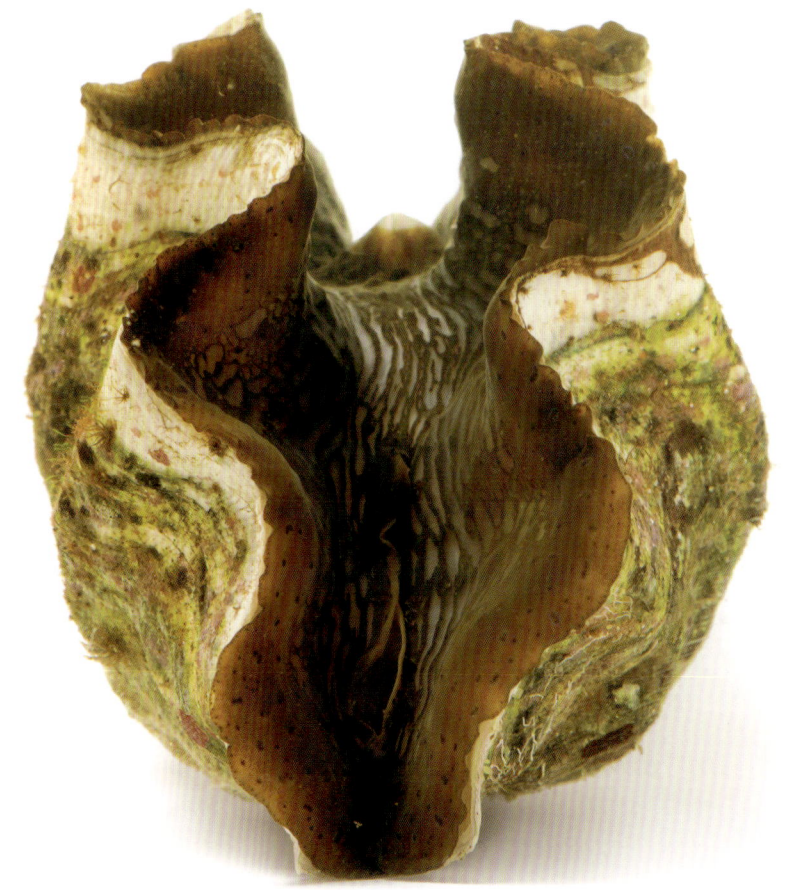

Giant clams

Tridacna gigas (VU)

The world's largest living bivalve mollusk, the giant clam—shown here as a juvenile—releases eggs and sperm into the ocean. After they unite, the developing larvae float in the sea, eventually attaching to sand or coral to continue growing.

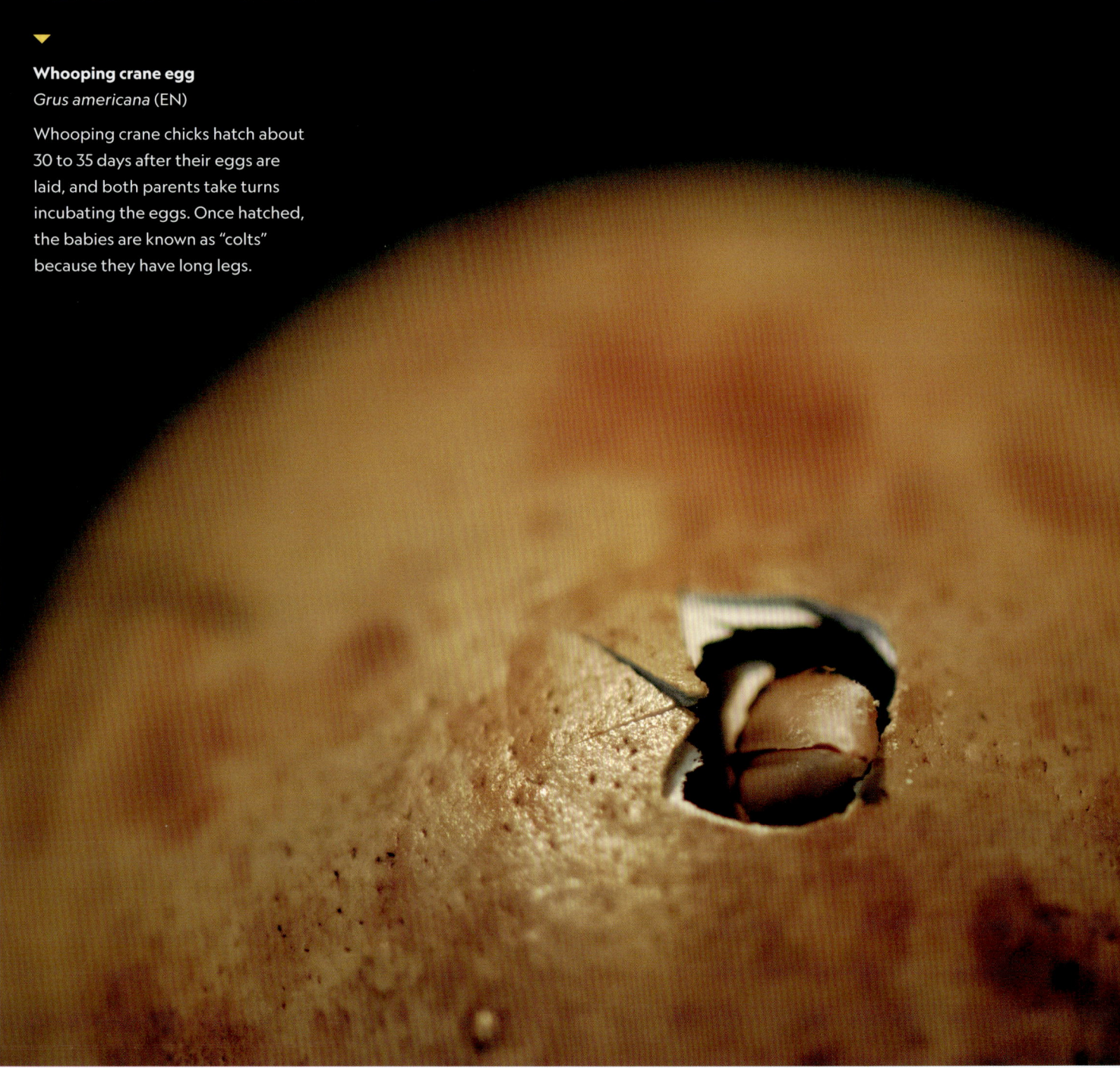

Whooping crane egg

Grus americana (EN)

Whooping crane chicks hatch about 30 to 35 days after their eggs are laid, and both parents take turns incubating the eggs. Once hatched, the babies are known as "colts" because they have long legs.

BRILLIANT DISGUISE

In the mid-1940s, there were only 21 whooping cranes left in the wild. Wildlife biologists decided the only way forward was to try to breed the birds in captivity—or else the species would go extinct. By the mid-1970s, captive cranes at Patuxent Research Refuge in Maryland had produced eggs, and over time, the captive flock grew larger.

For the last several decades, biologists with the U.S. Fish & Wildlife Service, the International Crane Foundation, and other agencies have learned the dos and don'ts of raising cranes in captivity: They know a crane pair needs lots of time to get to know each other, and that female cranes produce eggs only when they have a strong bond with their partner. And they know that crane chicks growing up in captivity run the risk of getting attached to their human caretakers. Young birds imprint on the first large moving object they see and assign that creature the role of parent. So whenever human caretakers interact with whooping crane chicks, they dress in head-to-toe crane costumes. The techniques appear to be working: Today, there are more than 800 whooping cranes worldwide—still a tiny number, but far better than 21.

66

"We need the same things that animals do: clean air, clean water, stable climate. Saving nature needs to be intentional now."

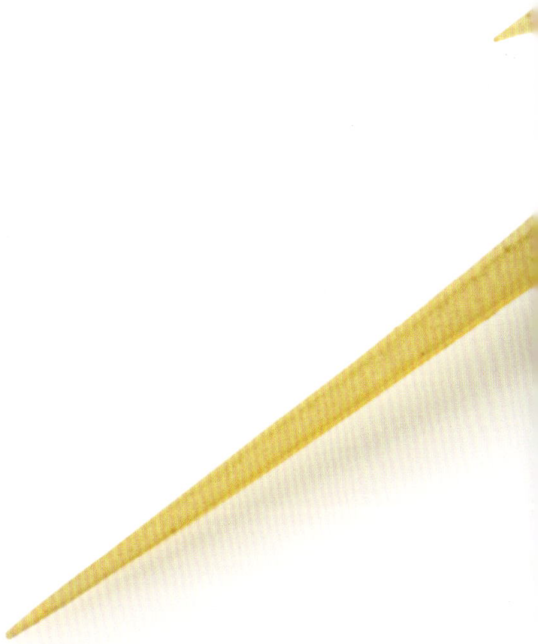

Atlantic horseshoe crab

Limulus polyphemus (VU)

Larval crabs hatch looking much like miniature adults. The young crabs head to sandy-bottomed tidal flats to continue their growth. Over the next 10 years, they'll molt 16 or 17 times before reaching maturity.

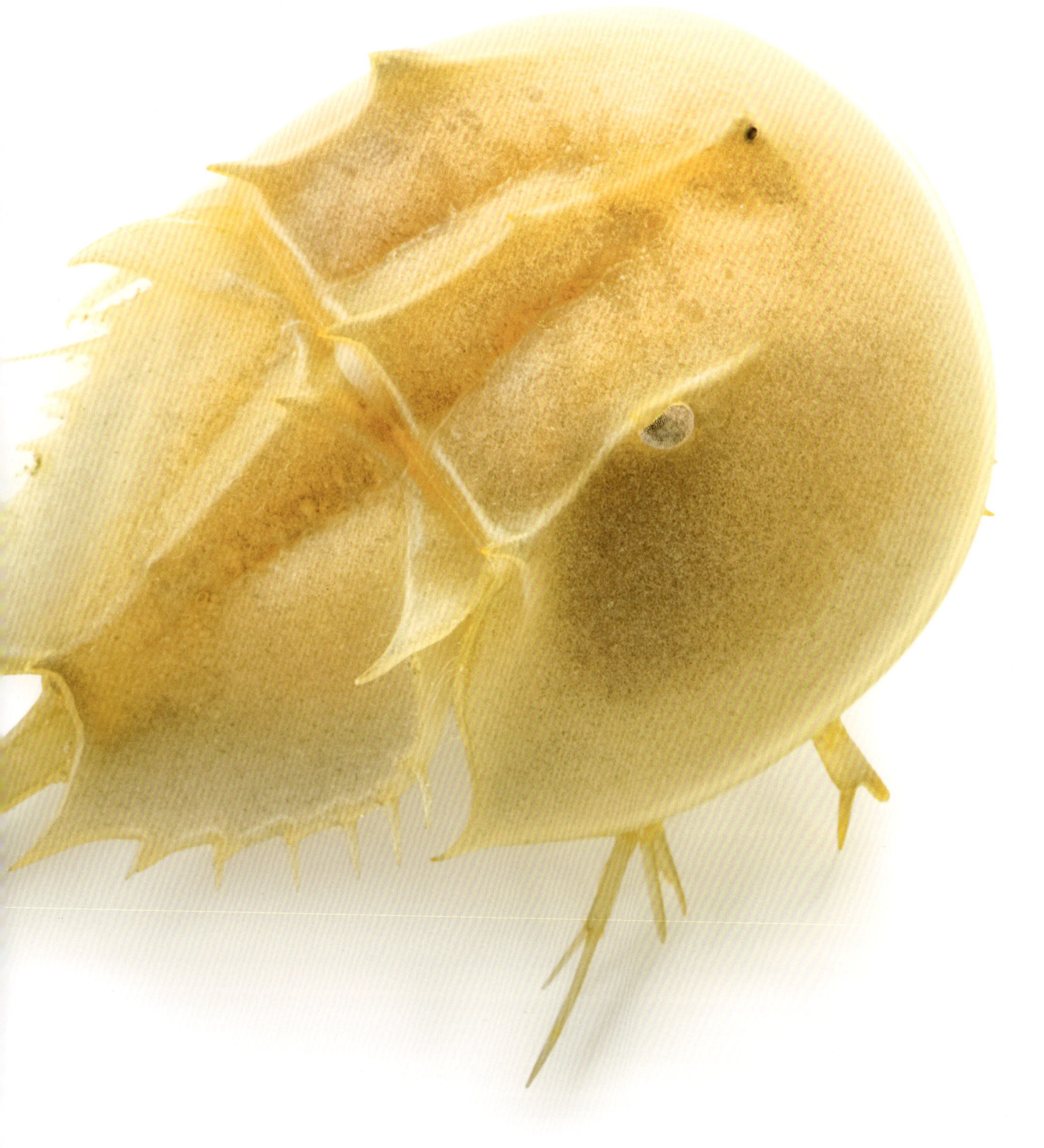

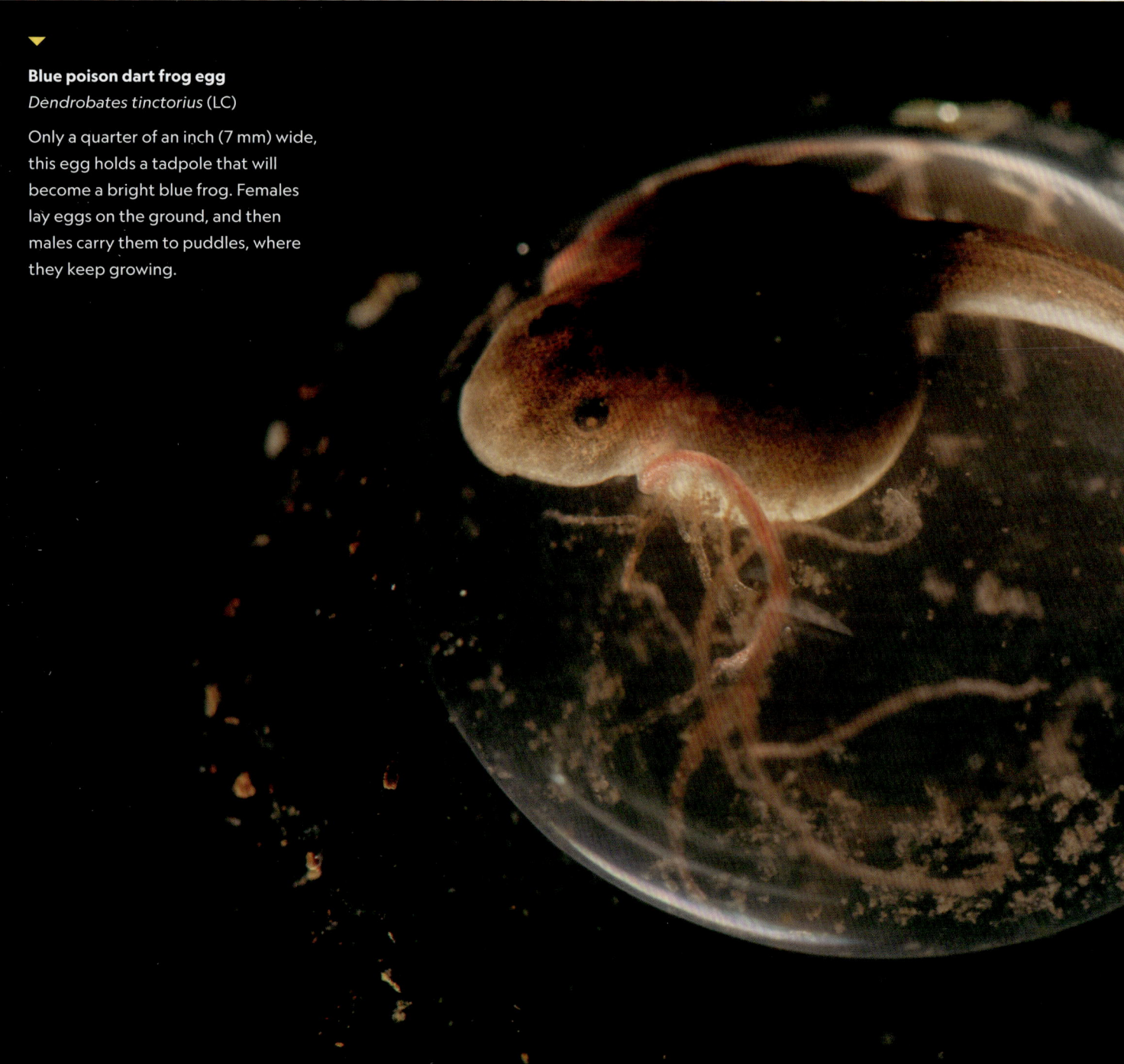

Blue poison dart frog egg
Dendrobates tinctorius (LC)

Only a quarter of an inch (7 mm) wide, this egg holds a tadpole that will become a bright blue frog. Females lay eggs on the ground, and then males carry them to puddles, where they keep growing.

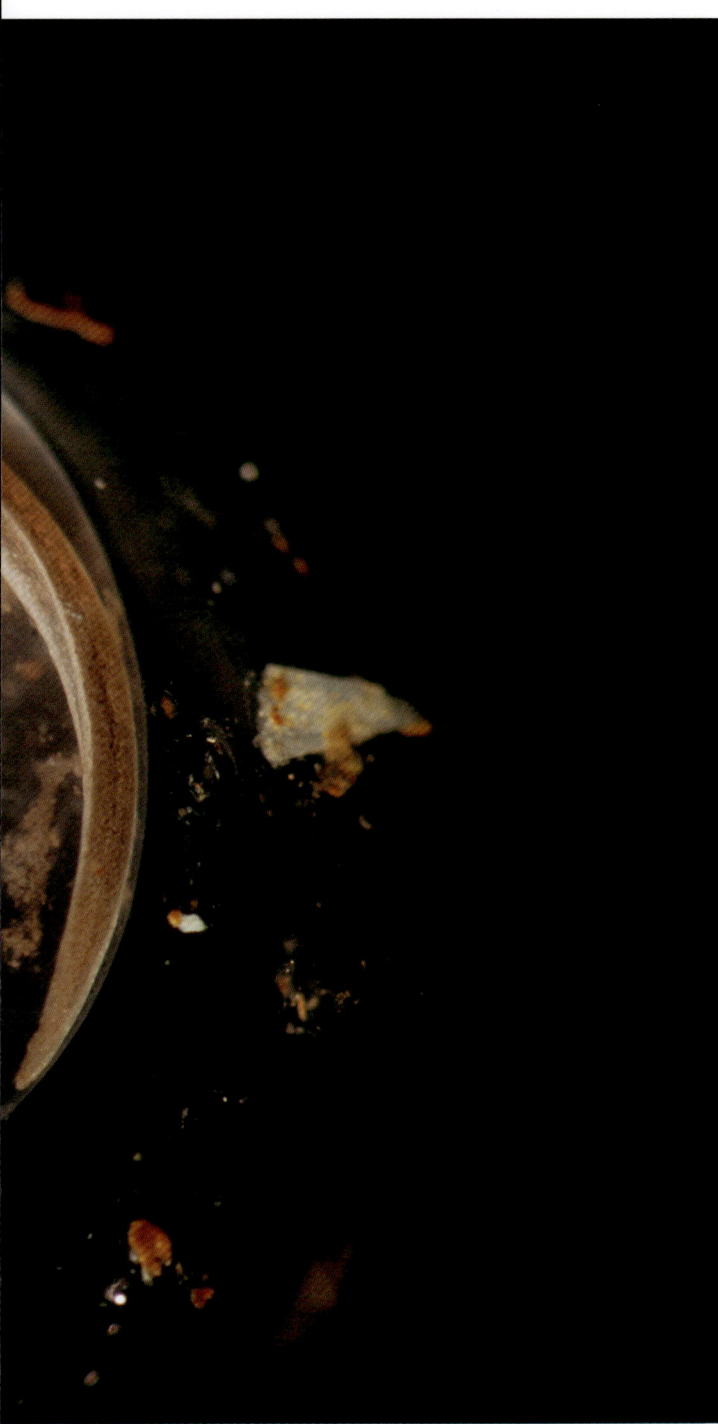

BEHIND THE LENS

HANDLE WITH CARE

ANIMAL: Blue poison dart frog egg, *Dendrobates tinctorius* (LC)

LOCATION: Private collection

"It was an overcast, rainy day, and I was at the home of a man who was a successful breeder of poison dart frogs. On misty, wet days, he would open up the windows in his spare bedroom, where he kept the frogs. They sensed the humidity and would start calling. He was able to mimic nature and get his frogs to breed. This particular egg was smaller than a pea and incredibly fragile. We couldn't just lay it out on a board or put it in a photo tank; we couldn't let the egg dry out. So we put the egg in water in a little petri dish over a black background, photographed it quickly, and put it back. These frogs have been around for millions of years, but they're so delicate, especially in the beginning. This picture reminds us how fragile they are."

GOTCHA
Take the Bait

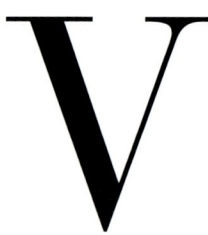

VERY FEW ANIMALS have an origin story quite like freshwater mussels. Males release sperm into the water. Females catch as much of it as they can using their siphons—the body part through which they filter water—to fertilize their eggs. The eggs hatch within the female's shell. Then comes the tricky part: The mussel larvae are tiny, shell-less, and vulnerable. They won't survive in the open water without some help from fish.

Different species have different ways of accomplishing this: Some mussels extend a fleshy protuberance into the water. Waving and wiggling in the current, it looks like a fishing lure. Attached to one end is a tiny sac of mussel larvae: a big bite for a passing fish to gulp. The larvae head for the fish's gills, where they attach and remain until they're big enough to release themselves into the water.

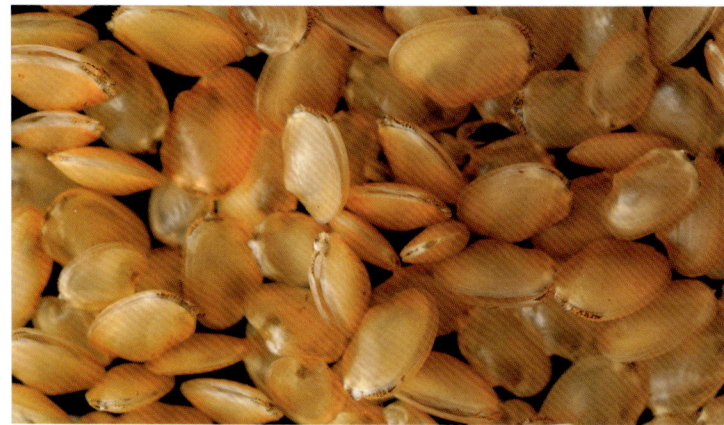

Pink mucket mussels *Lampsilis abrupta* (VU)

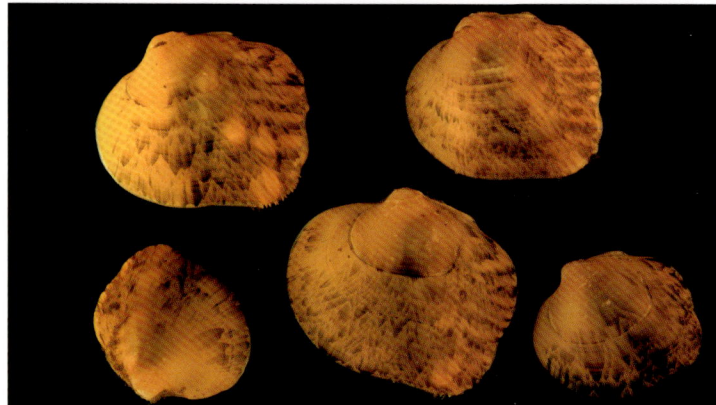

Appalachian monkeyface mussels *Theliderma sparsa* (CR)

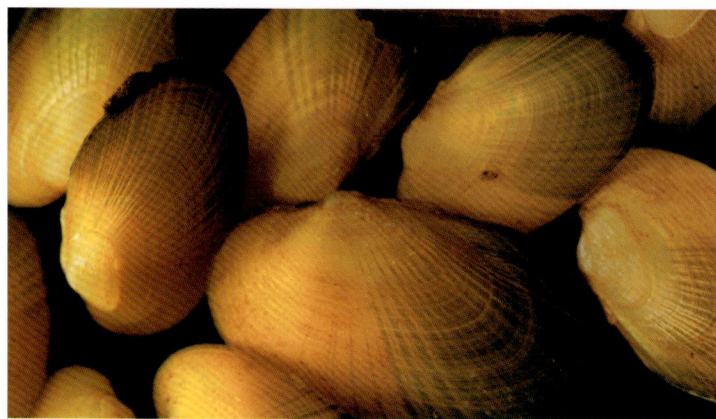

Rainbow mussels *Villosa iris* (NE)

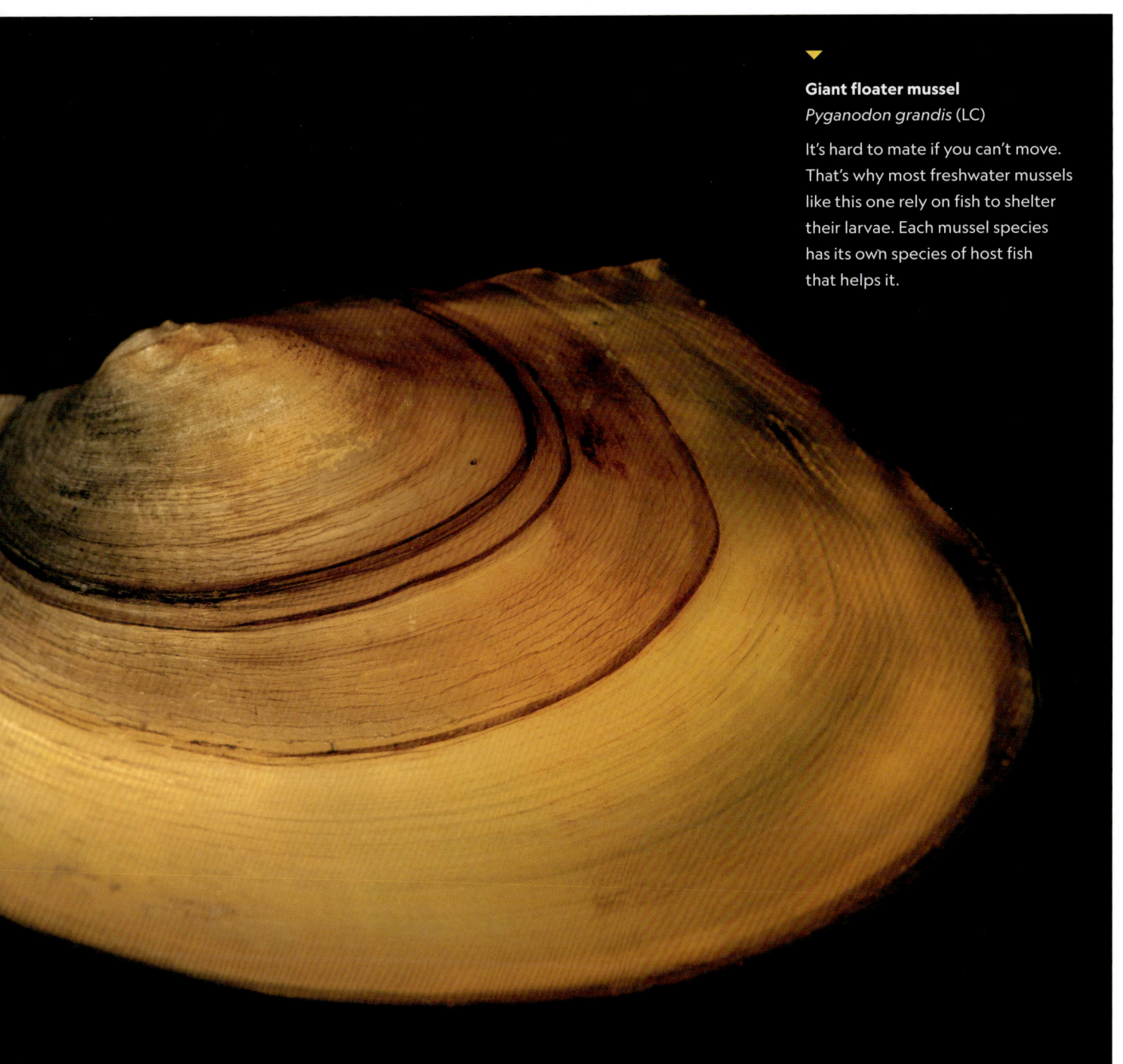

Giant floater mussel
Pyganodon grandis (LC)

It's hard to mate if you can't move. That's why most freshwater mussels like this one rely on fish to shelter their larvae. Each mussel species has its own species of host fish that helps it.

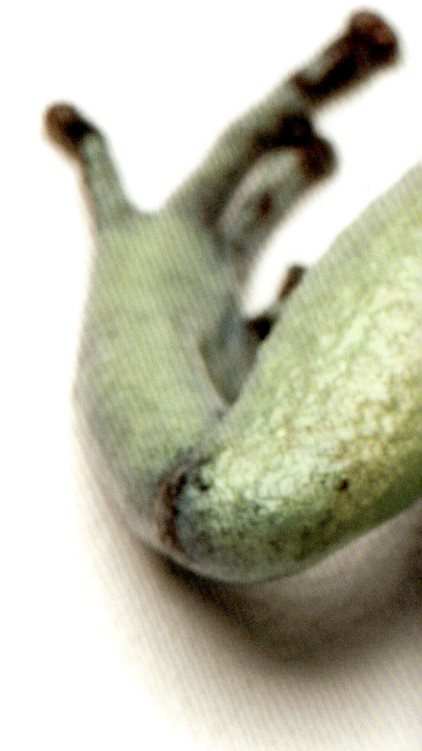

2 | NEWBORN

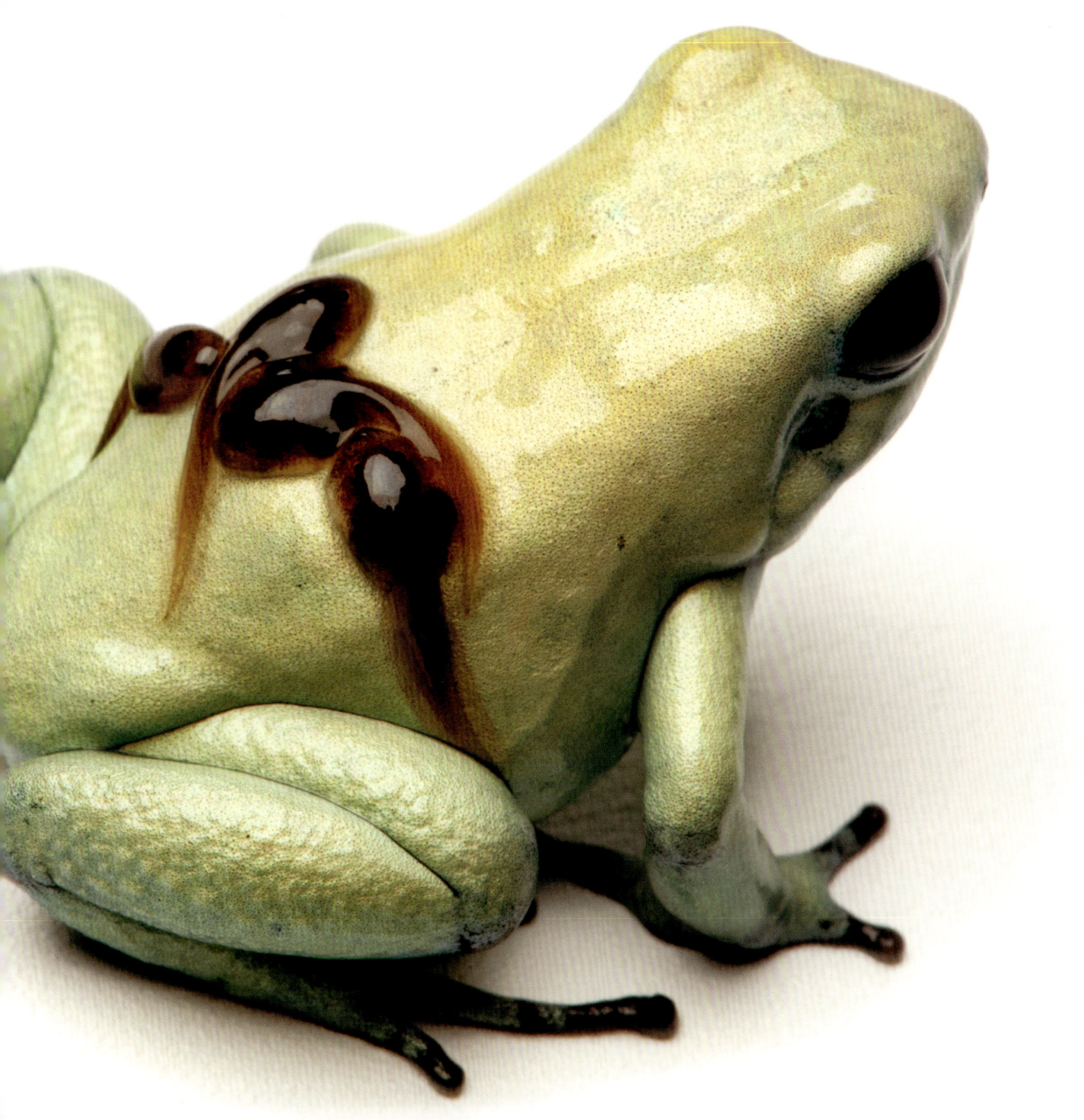

Welcome to the World

Baby animals are so delightful that we surround
our own infants with their likenesses. These youngest of animals are evolutionary
miracles as well: Newborns know just what to do.

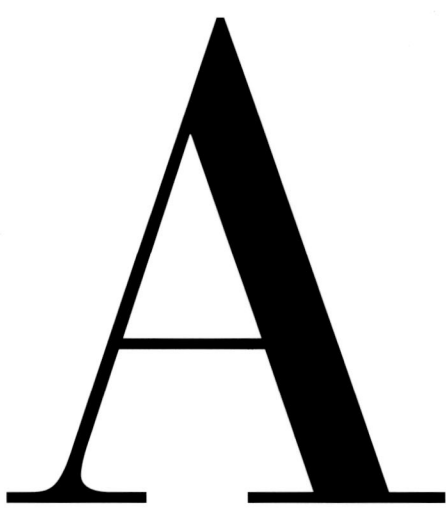

A BABY ANTELOPE PUSHES its way into the world, stands on wobbly legs, and within hours is ready to run with the herd. A newborn wallaby no bigger than a bean burrows deep inside its mother's warm pouch, not yet ready to leave. Least tern chicks can walk and swim soon after hatching; on toothpick legs, they scurry for shade and shelter.

For animals of all kinds, the newborn stage marks the starting line in a race to adulthood. But that's about where the similarities end. Some animals, like the hoofed mammals of the open plains and rugged mountains, set off with a bang. They are born as tiny models of their parents, fully developed and soon ready to move. They have to keep up with the herd from the first days of their lives.

Others, born or hatched helpless and weak, bide their time in the protection of their parents, who feed and care for them until they are bigger and stronger. Insects are an altogether different story. Shape-shifters from the start, insects like ants, bees, beetles, wasps, and moths hatch as larvae, and will not resemble the adults of their species until they complete a full metamorphosis. Others, such as grasshoppers, dragonflies, and mantises, hatch as tiny versions of adults, then go through several stages of molting before they reach adult size.

Newborn bodies are brand-new and vulnerable. Baby animals may be covered in down or fur for warmth, or have scales, spines, or a hard shell for protection. Spots, speckles, and stripes create camouflage to hide the newborns on the forest floor or in the coral reef. Some greet the world in sheltered environments like shallow lagoons or deep burrows—nurseries of a sort where conditions are mild and predators scarce.

The newborn stage is so critical to social animals that if anything goes awry, the repercussions can persist into adulthood. Many species of monkeys cling to their mothers' bellies or backs, feeling safe only there. Scientists have found that babies of some species who cannot be with their mothers in this way during early life will experience troubled behavior as adults.

In many ways, newborns of the animal kingdom are not so different from our own. They eat a lot. They sleep a lot. They grow and change by the minute, at once both helpless and invincible.

And when animal babies look about with wide, alert eyes, we feel somehow they are capable of seeing in a way that we cannot. What must they think of their strange and wondrous new world?

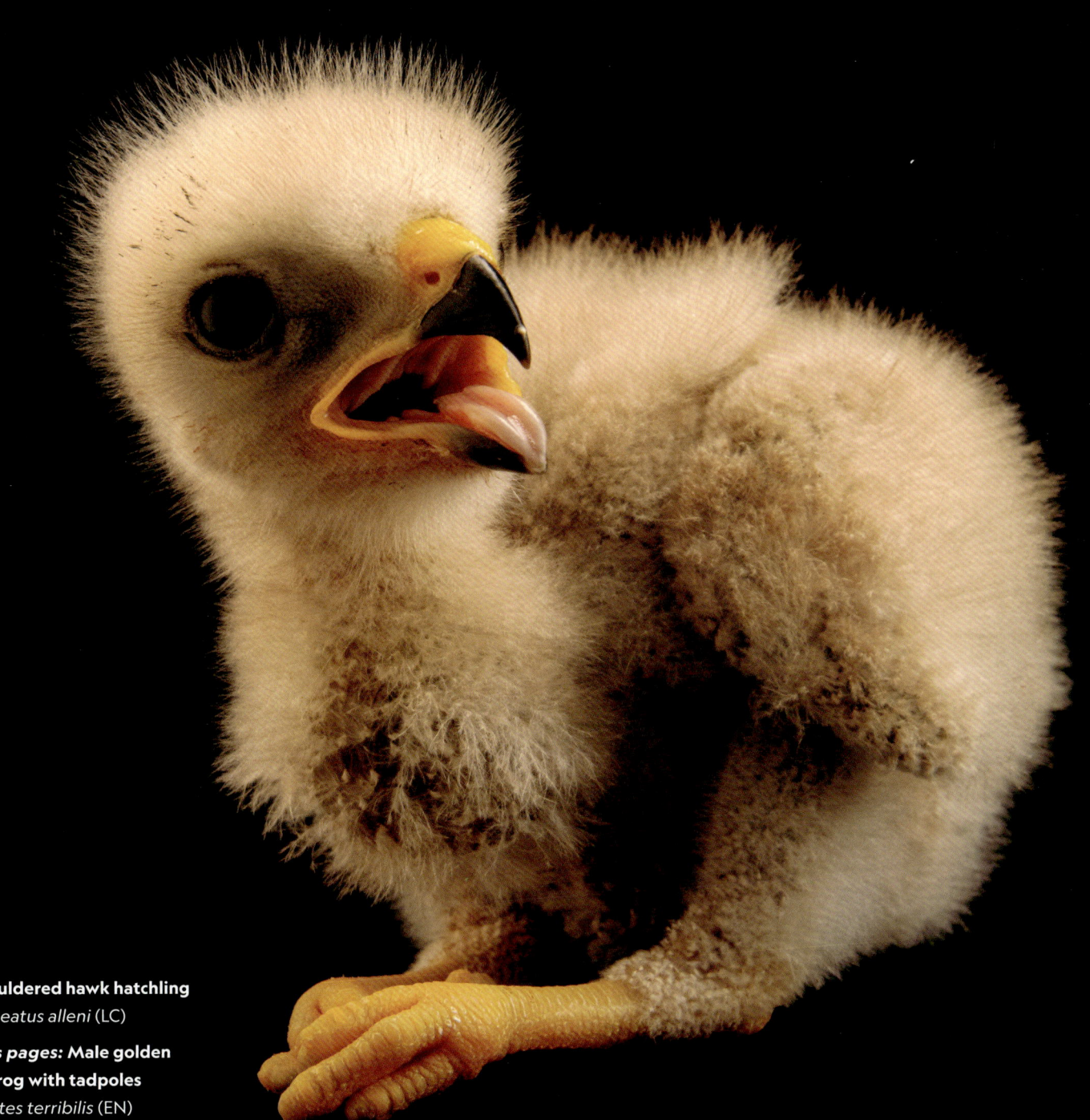

Red-shouldered hawk hatchling
Buteo lineatus alleni (LC)

***Previous pages:* Male golden
poison frog with tadpoles**
Phyllobates terribilis (EN)

SURPRISE!
Winter Babies

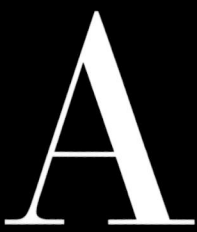

ADULT BLACK BEARS mate during June and July. A fertilized egg within the mother's belly develops into a small cluster of cells. If the female gains enough weight by late fall, those cells will survive and the embryos will grow. The mother bear gives birth in January or February during the first month or two of hibernation, a period of time when bears' metabolic and physical activities are low.

Naked and blind, the little cubs nurse from their lethargic mother for weeks. By the spring, they are much bigger, weighing between four and 11 pounds (2 and 5 kg). Bear cubs stay with their mother for about 18 months. From her, they learn how to be bears: how to forage for food and find their way through their environs. After their second winter together, when the mother is ready to mate again, her youngsters take off on their own.

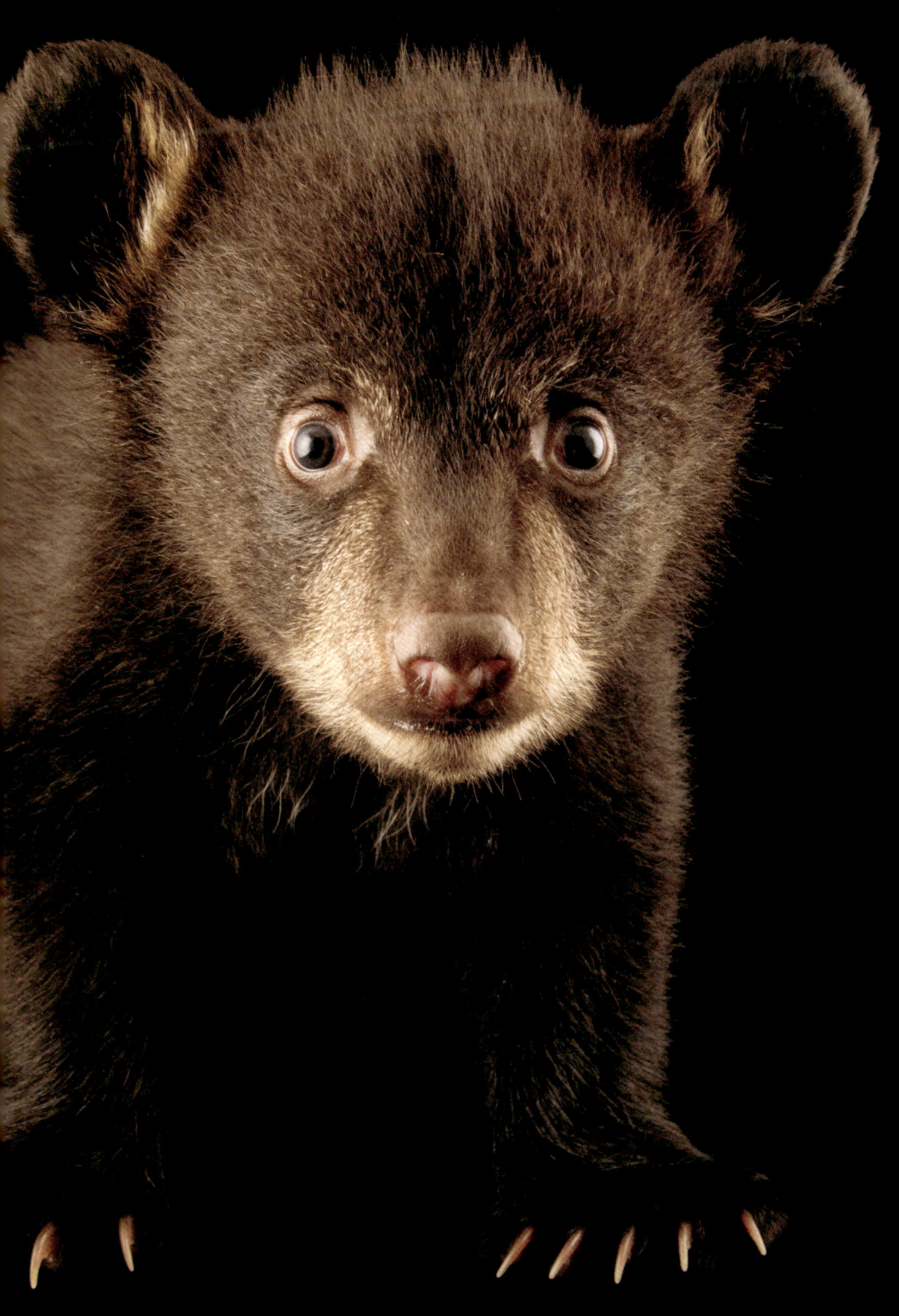

▼

Eastern black bear
Ursus americanus americanus (LC)

Usually born in litters of two or three, bear cubs start out tiny, each weighing a pound or less (between 200 and 450 g). They are the smallest babies relative to the adult's size of any mammal.

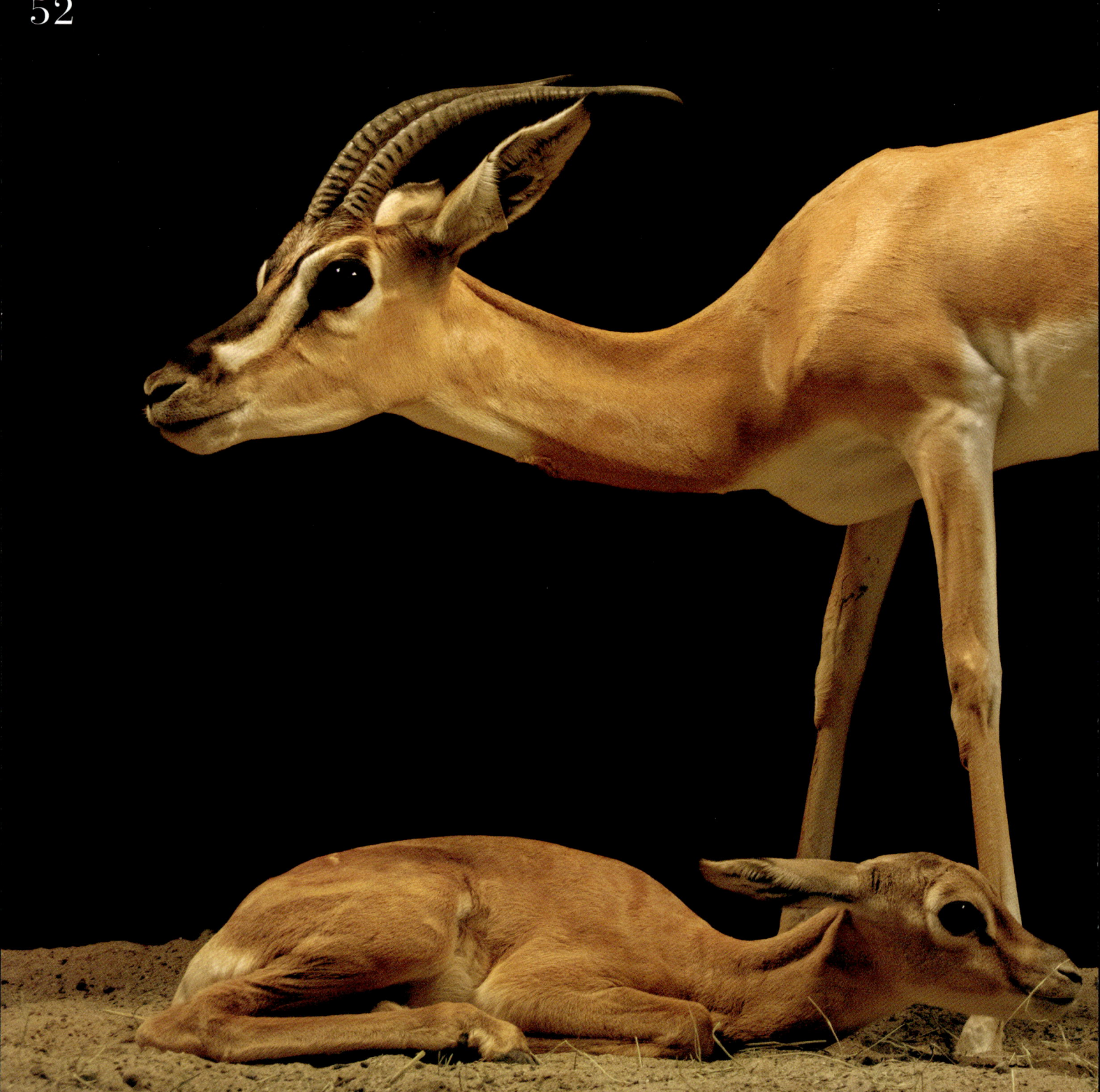

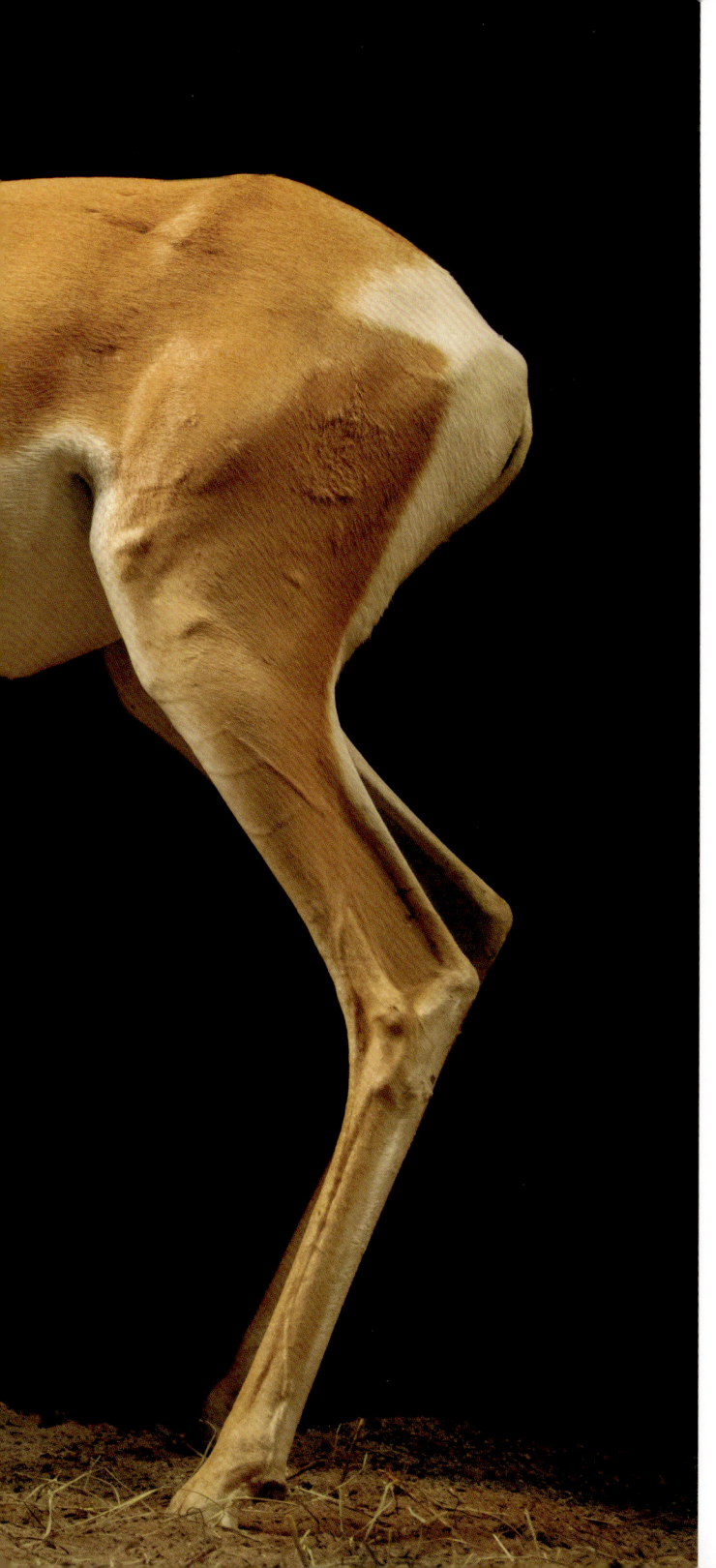

▼

Soemmerring's gazelles

Nanger soemmerringii (VU)

Female gazelles like this one, pictured here with her calf, roam the grasslands and scrublands of the Horn of Africa in herds of 15 to 20 other females. They typically give birth to one calf a year.

Clouded leopards

Neofelis nebulosa nebulosa (VU)

At nine days old (left), leopard kittens' eyes are still shut, and they can't yet walk. By nine weeks (right), they can walk and even climb trees—an important skill for a stealthy forest hunter.

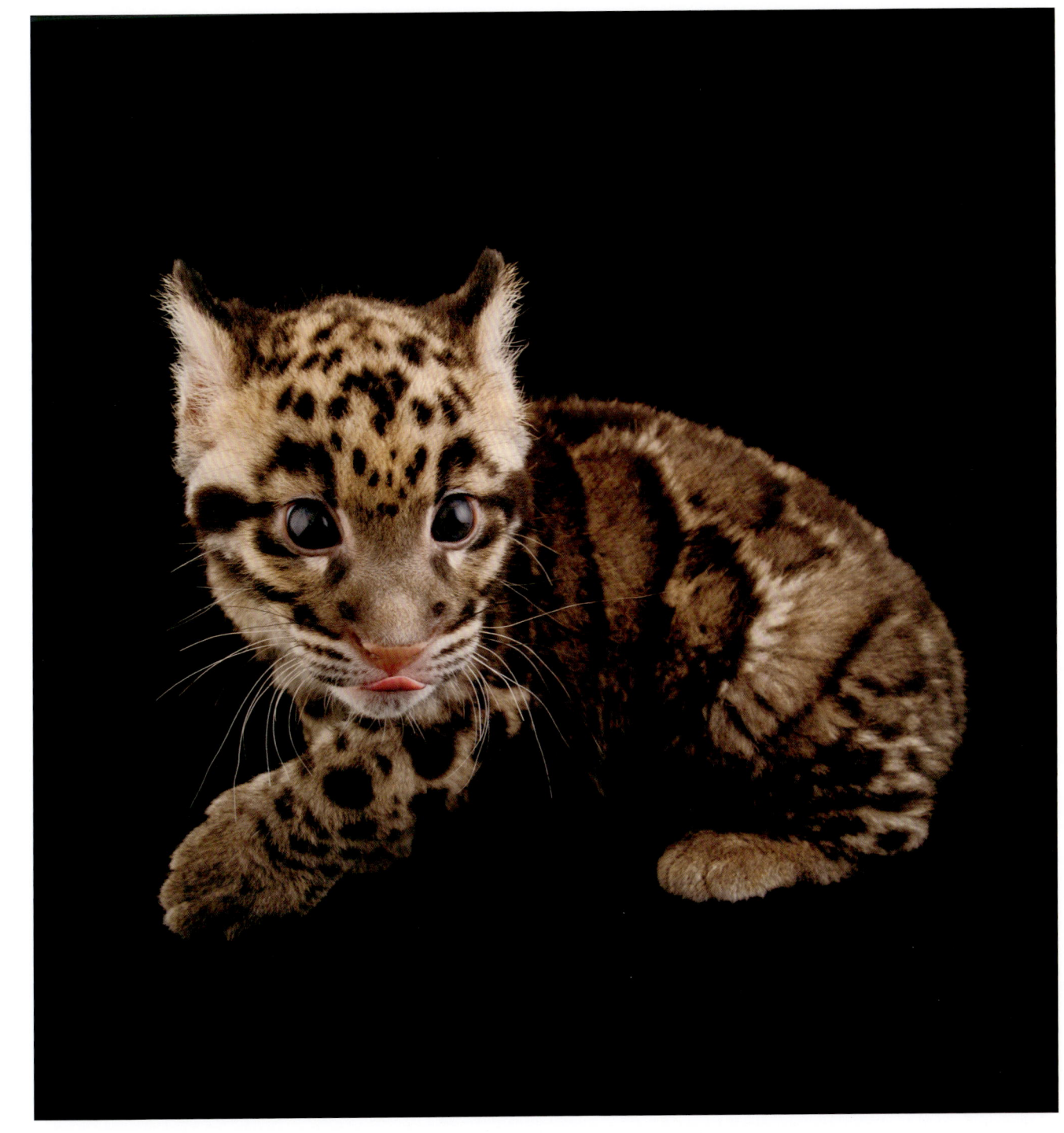

Life With Mom

Stay Close

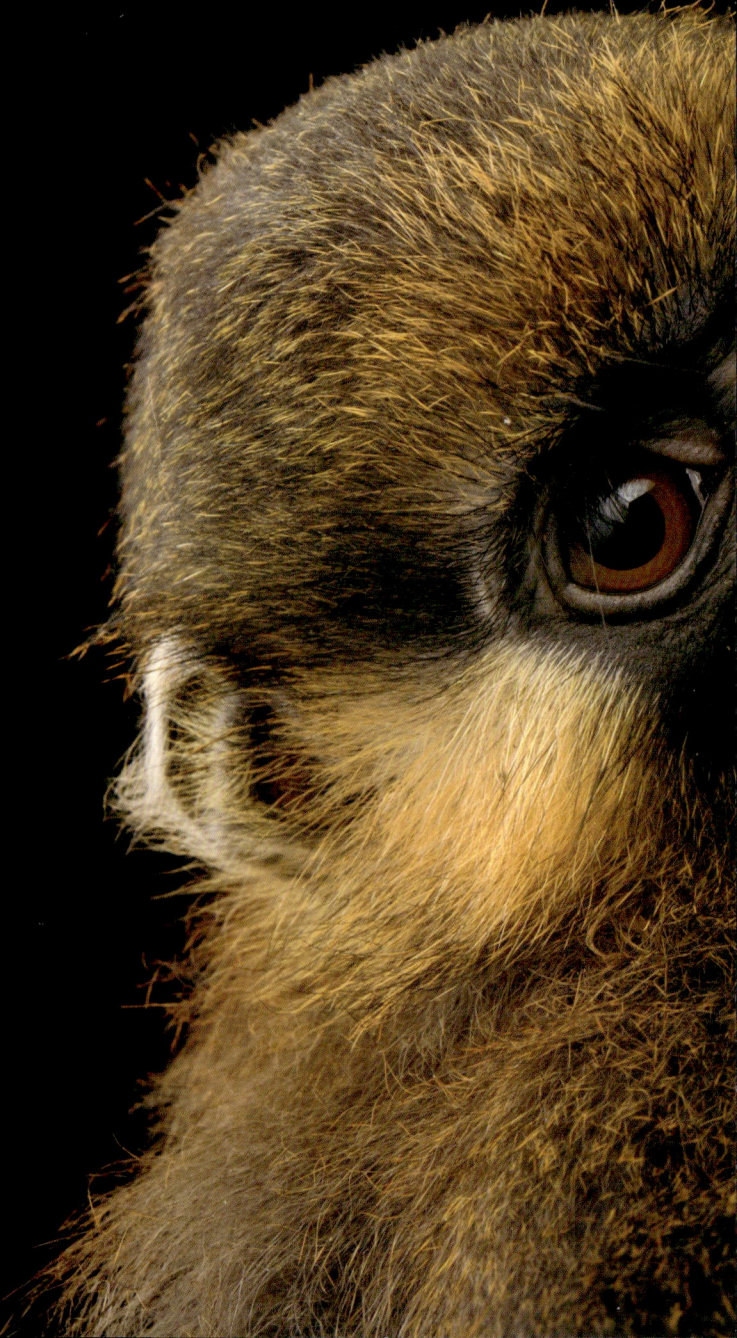

T

THE AGILE gray-tailed moustached monkey of the tropical forests of western Africa is born relatively well developed. At birth, a baby moustached monkey can open its eyes, support its own weight, and hug its mother. The monkeys live in groups of six to 10 individuals, and moms care for their babies for a full year. The young monkeys almost always cling to their mother's belly or back as she wanders the forest, leaping from tree to tree in search of fat-rich palm nuts, seeds, insects, and other food.

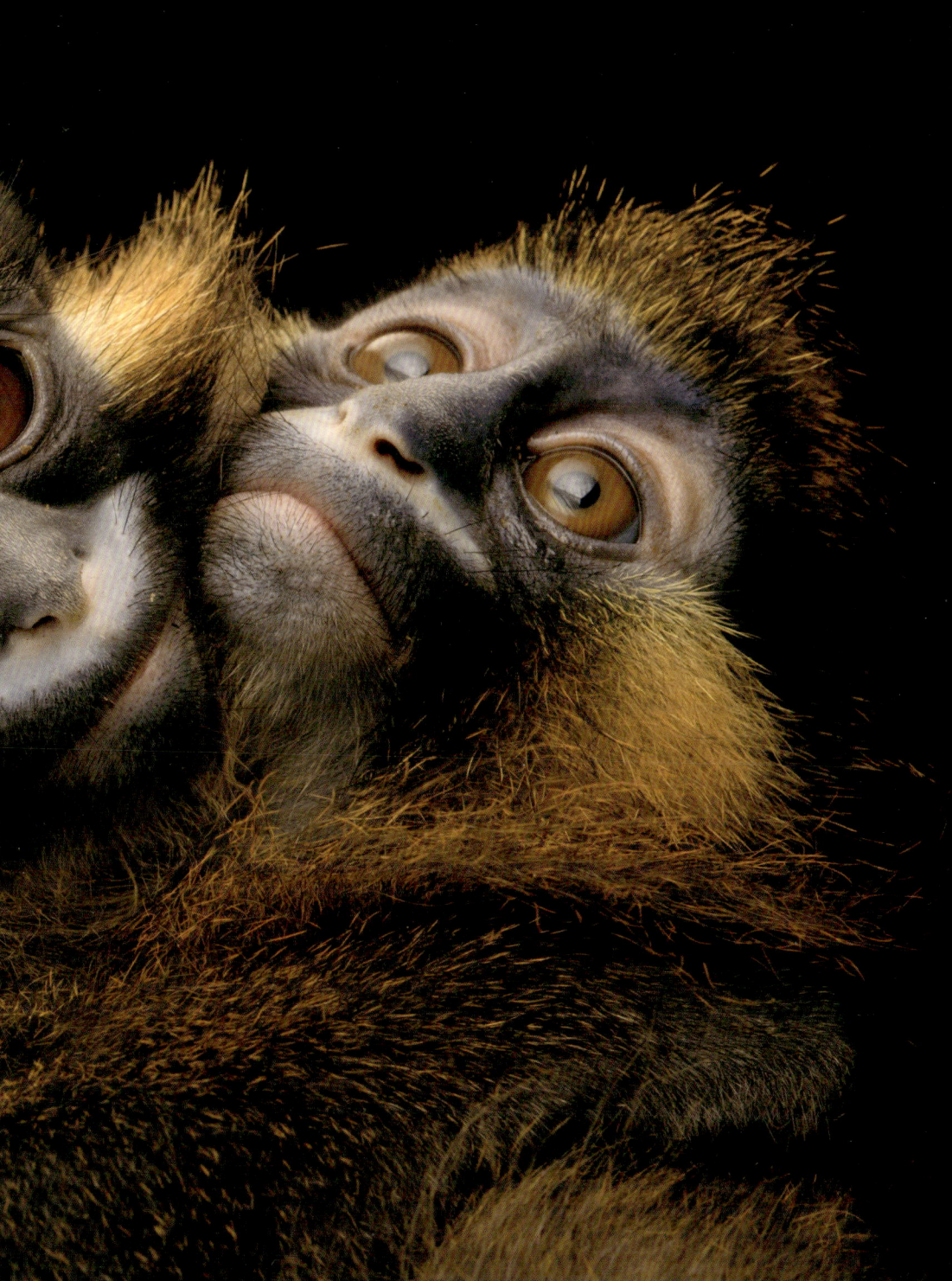

▼

Gray-tailed moustached monkeys
Cercopithecus cephus cephodes (NT)

Young monkeys are known to greet adults in their social group by emitting a soft, trilling call that oscillates in a descending pitch, signaling their obedience to their elders.

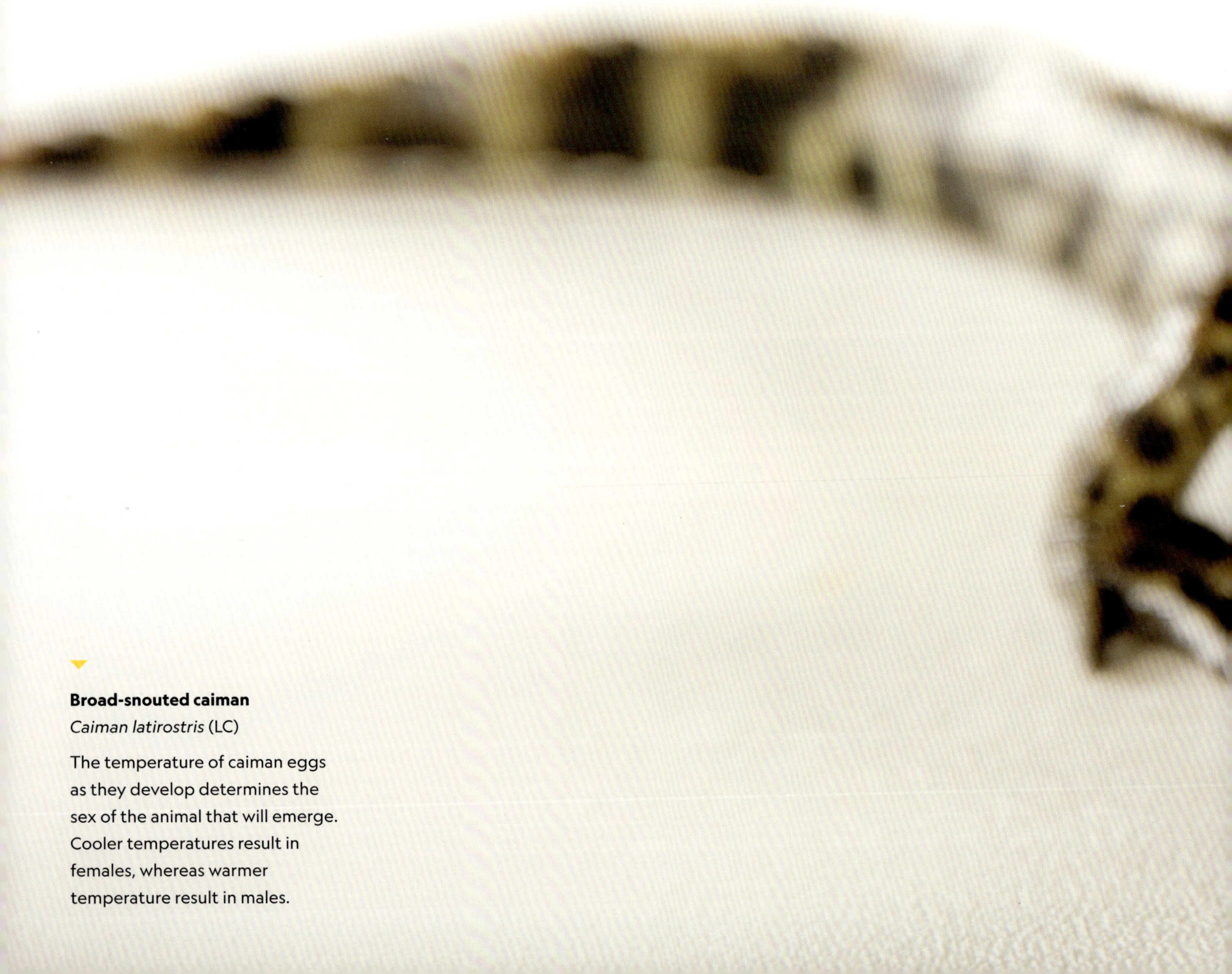

Broad-snouted caiman

Caiman latirostris (LC)

The temperature of caiman eggs
as they develop determines the
sex of the animal that will emerge.
Cooler temperatures result in
females, whereas warmer
temperature result in males.

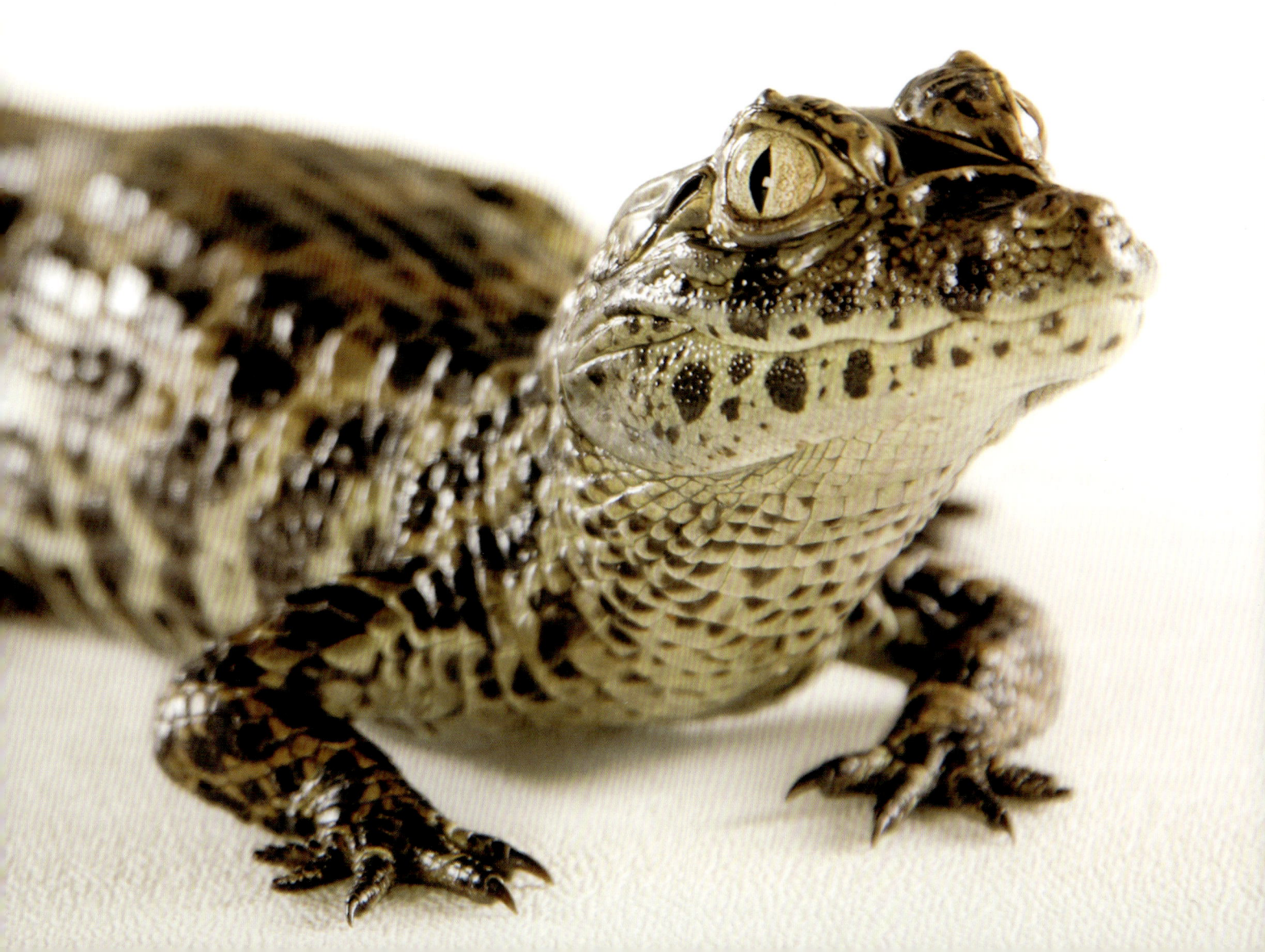

Free Fall
From Tree to Water

I

IN THE AMAZONIAN lowlands of South America, the fringed leaf frog, like others of its kind, has a particular approach to laying eggs. Eggs have to be moist, and they also need to be safe. While some species of tree frogs build a foam nest near water and lay their eggs in the protective foam, the fringed leaf frog descends from higher branches to lower ones to breed, then glues its eggs directly onto a leaf or twig near small pools of water—depressions in the ground, perhaps, or cavities in fallen logs. The eggs must stay above water while they develop; if they fall in, they'll perish. Once they hatch, though, the tadpoles immediately wiggle toward the nearby water, where they will continue to develop through their metamorphosis into frogs.

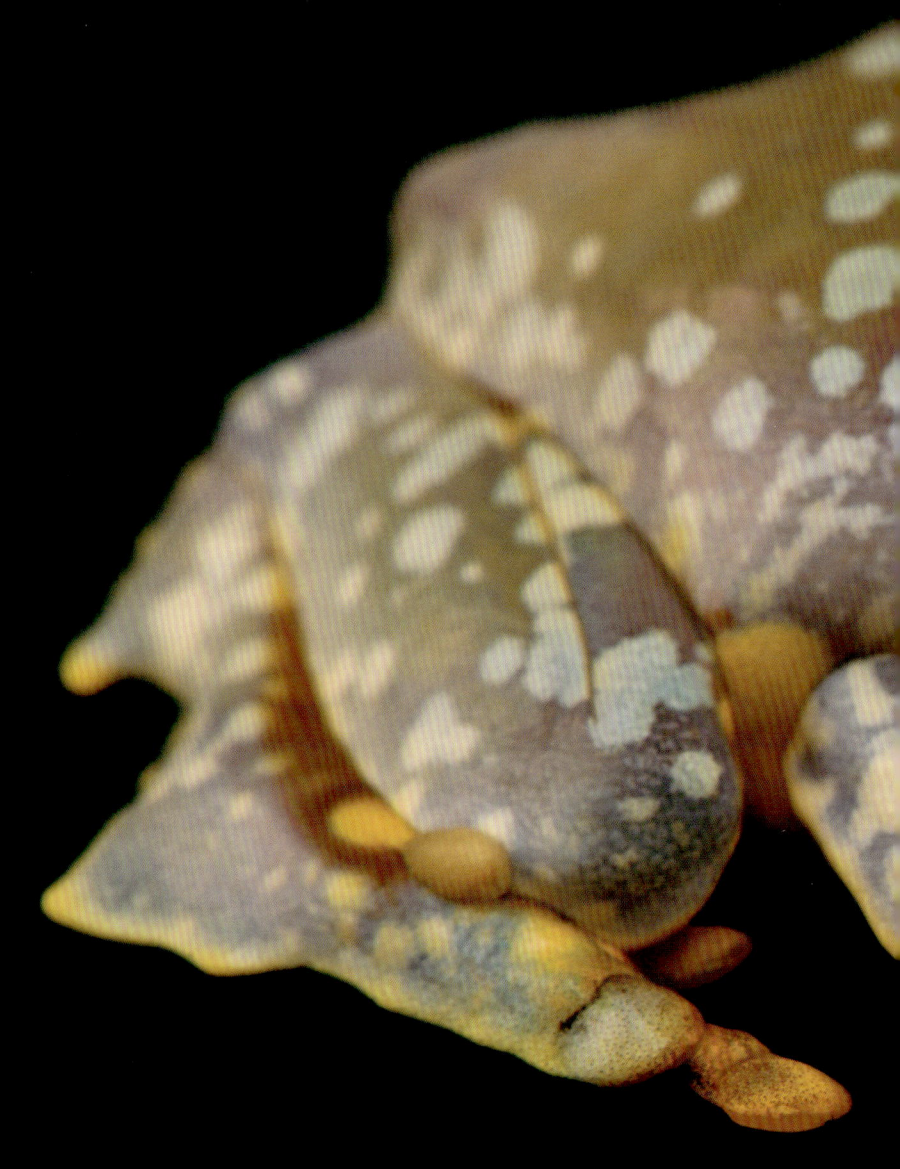

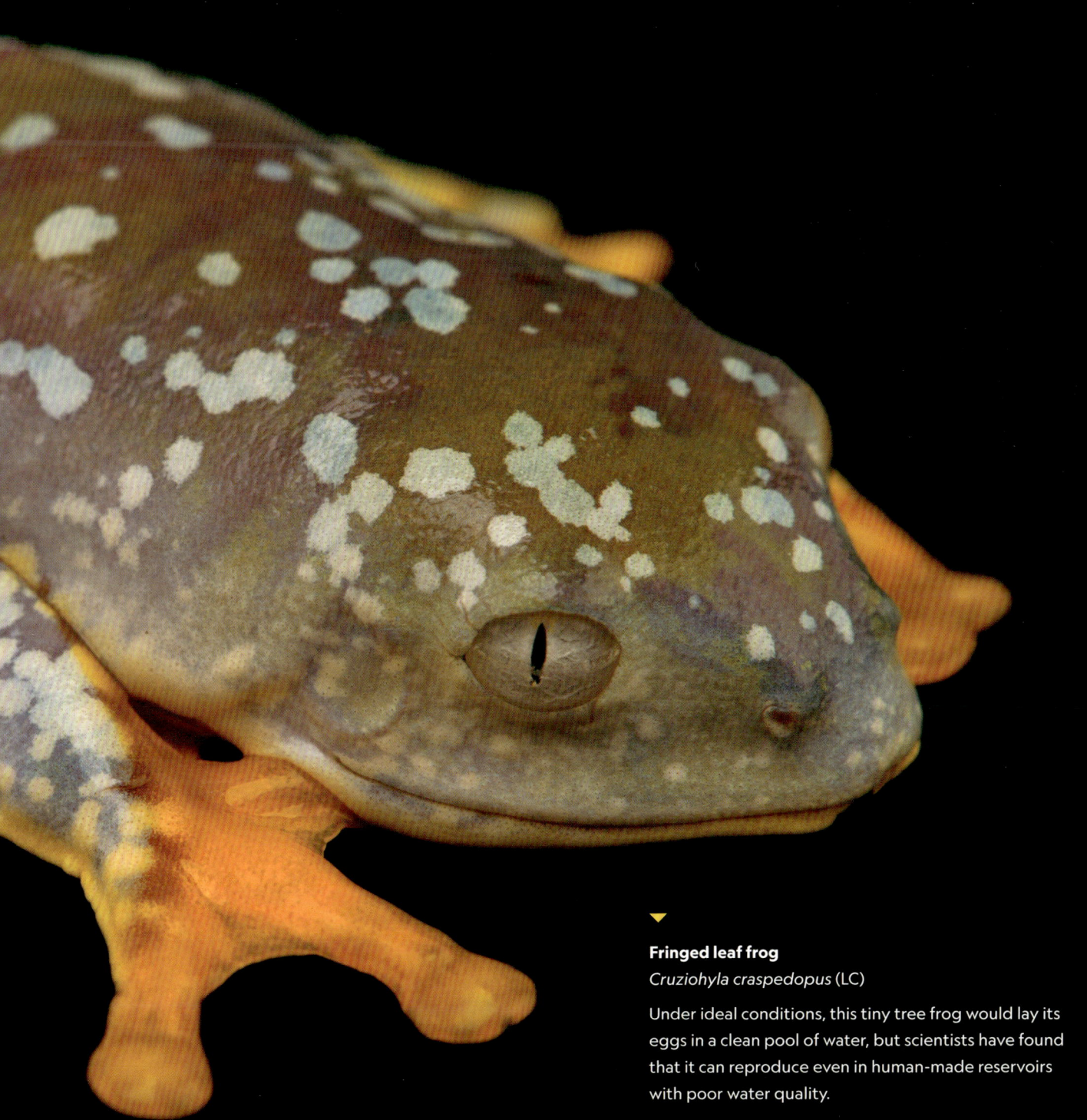

▼

Fringed leaf frog

Cruziohyla craspedopus (LC)

Under ideal conditions, this tiny tree frog would lay its eggs in a clean pool of water, but scientists have found that it can reproduce even in human-made reservoirs with poor water quality.

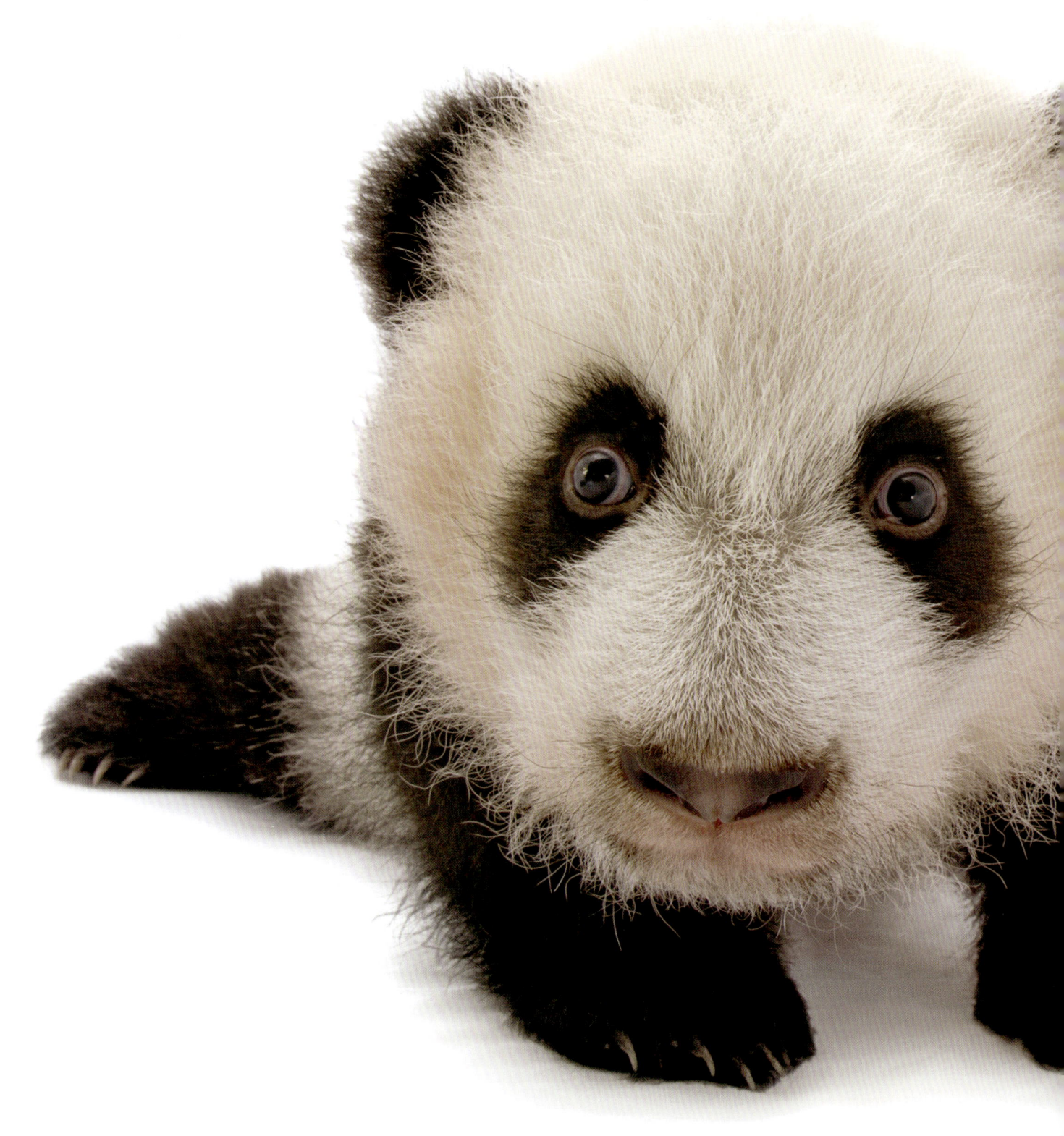

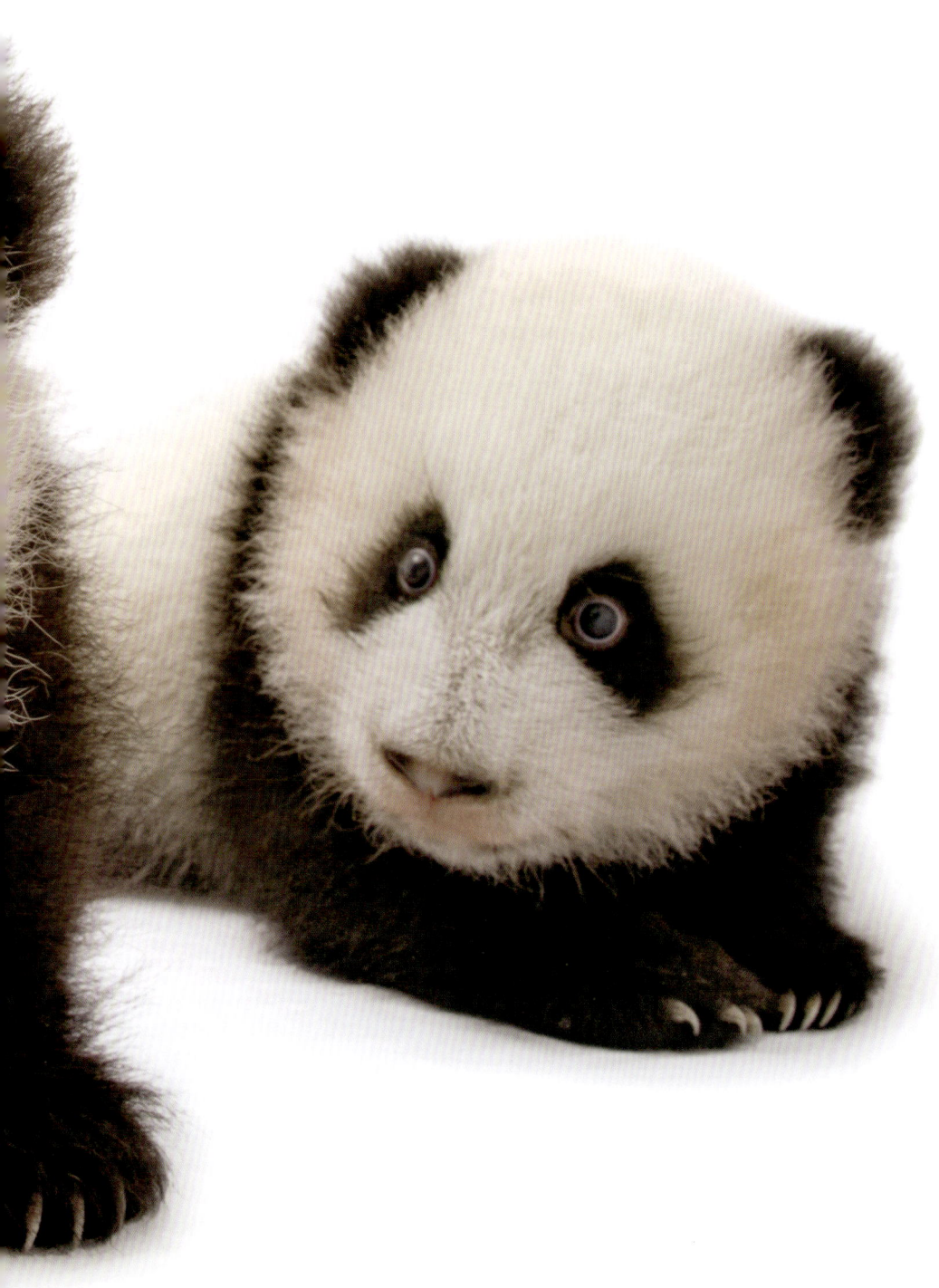

Giant pandas

Ailuropoda melanoleuca (VU)

When a cub is born, it is helpless and about the size of a stick of butter, even though its mother weighs roughly 1,000 times more than the cub. Newborns cannot open their eyes until they're three weeks old.

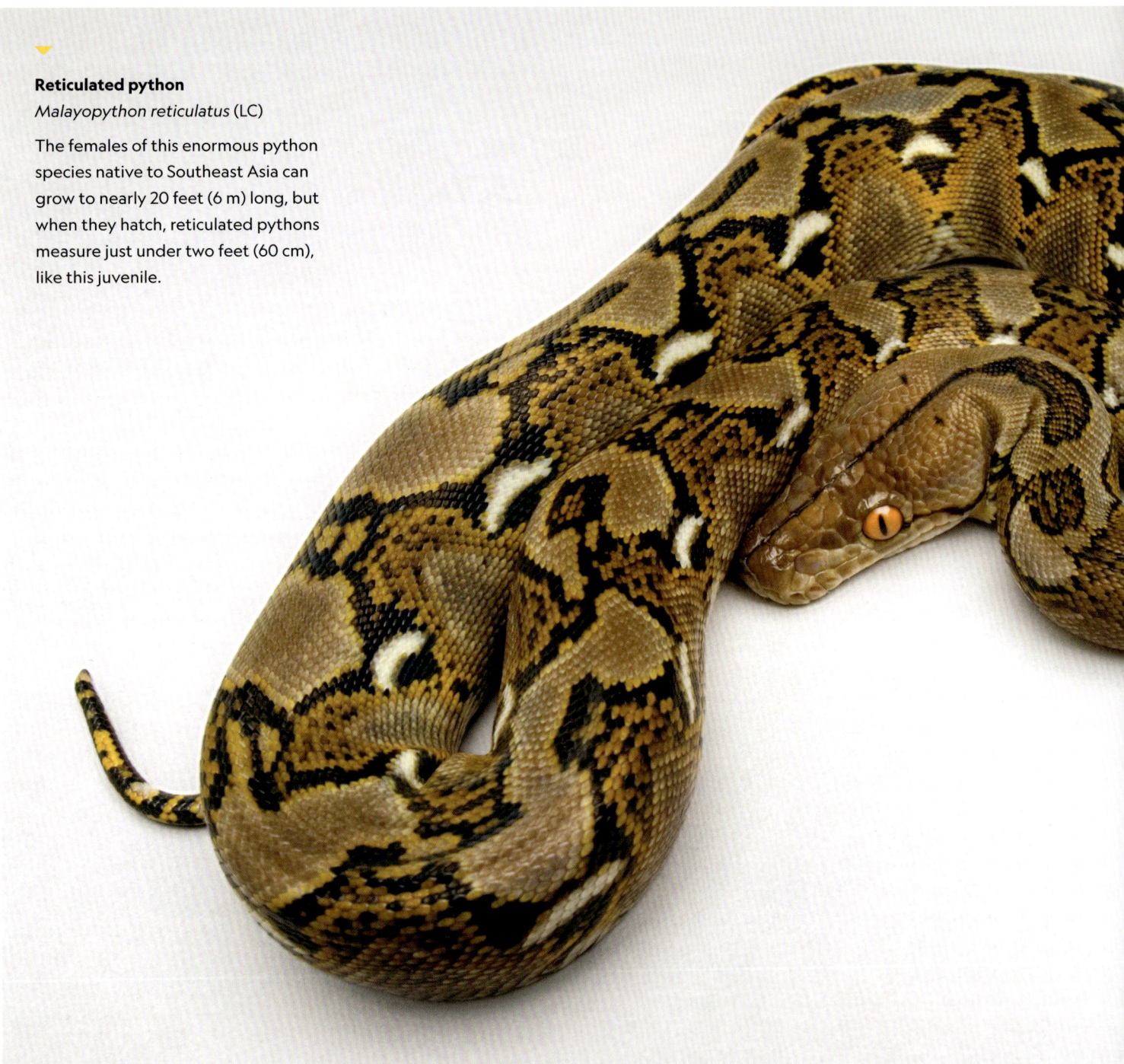

Reticulated python
Malayopython reticulatus (LC)

The females of this enormous python species native to Southeast Asia can grow to nearly 20 feet (6 m) long, but when they hatch, reticulated pythons measure just under two feet (60 cm), like this juvenile.

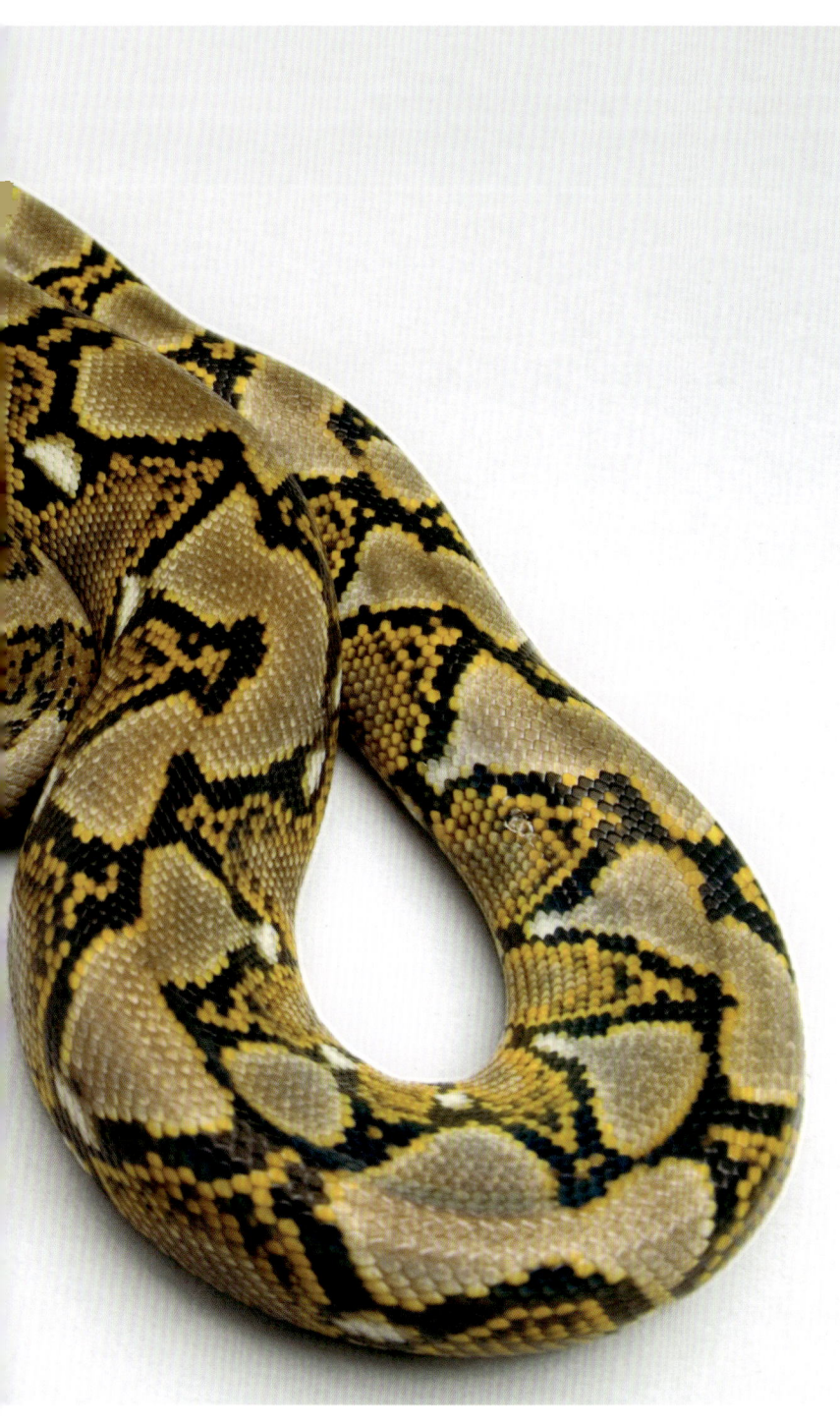

GENTLE GIANTS
Holding Still

T

THE FEMALES OF THIS enormous python species native to Southeast Asia may be huge, but they are caring mothers. For three months, female pythons remain coiled atop their clutch of eggs, which are usually laid in tropical rainforests, wetlands, or grassland forests, and almost always near water. The mothers do not eat, and they barely move. Slight muscle twitches are the only movement they may make; this raises the incubation temperature of their eggs.

When the baby snakes are ready to hatch, they use an egg tooth on their upper lip to break open the shell. In many other snake species, babies slither away from their mom quite soon after hatching. But young reticulated pythons may stick around for a few days before venturing off, basking in the sun on their mom's broad back.

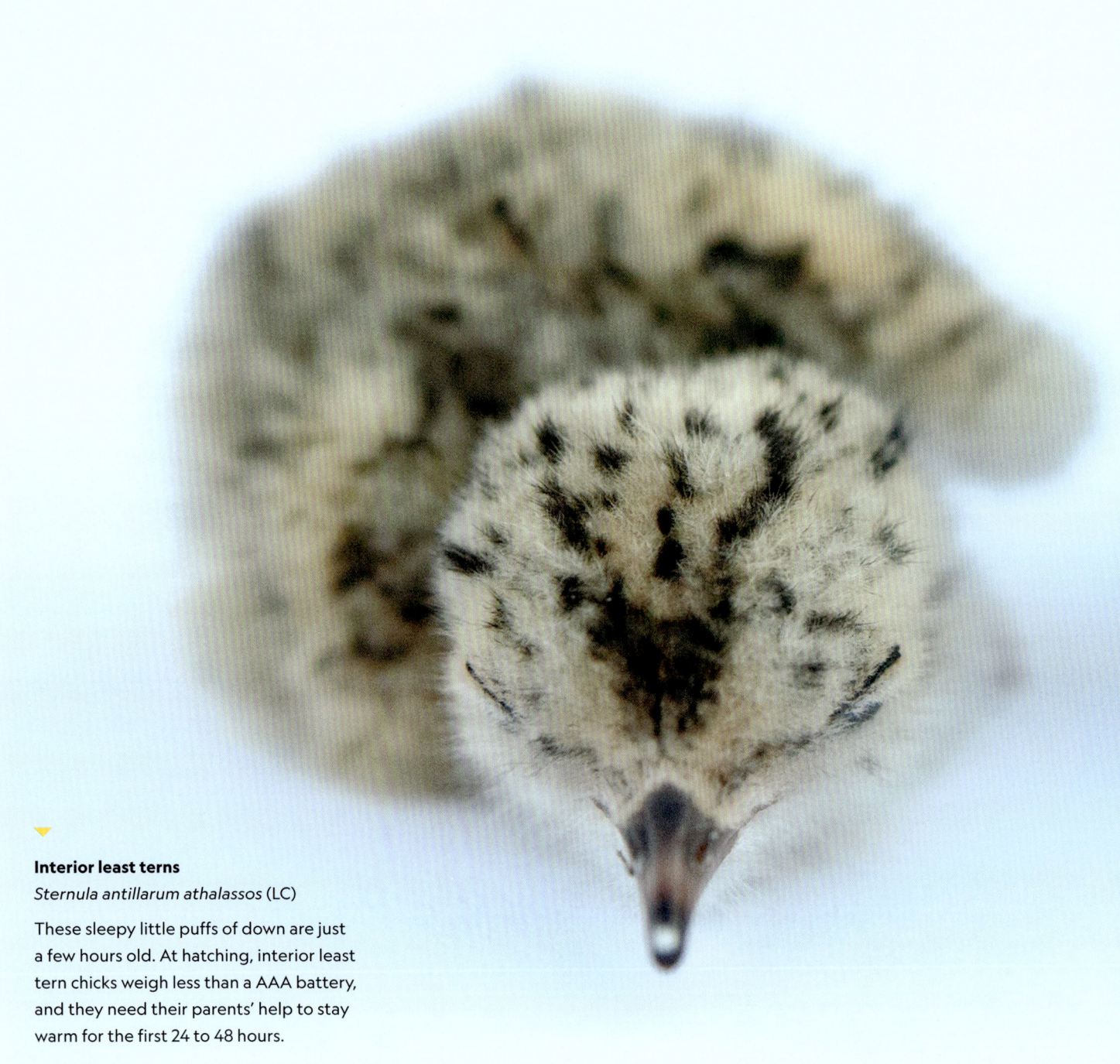

Interior least terns

Sternula antillarum athalassos (LC)

These sleepy little puffs of down are just a few hours old. At hatching, interior least tern chicks weigh less than a AAA battery, and they need their parents' help to stay warm for the first 24 to 48 hours.

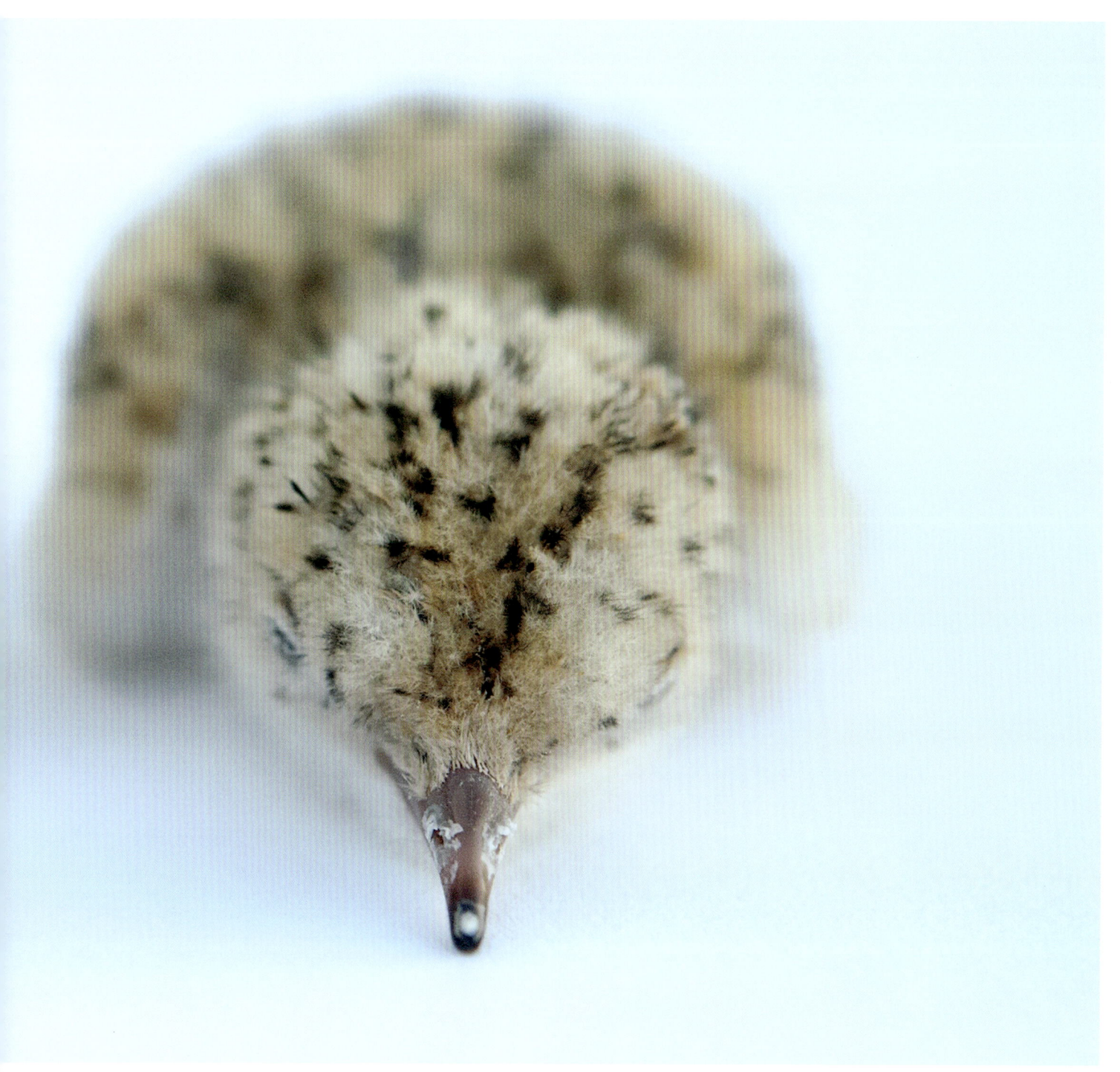

NIGHTLIFE
Rolling Along

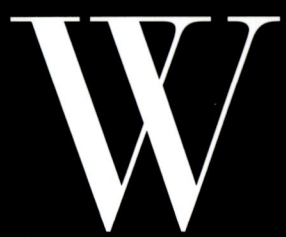

W

WHEN A MOTHER pangolin gives birth, her pango-pup is about six inches (15 cm) long and weighs less than a pound (0.4 kg). For about three months, a pangopup nurses from its mother. When it's a month old, it also starts eating the pangolin's preferred food: termites and ants.

The little white-bellied pangopup rides around on its mom's tail while she wanders the lowland forests and woodlands of Central and West Africa at night, foraging for insects with her long, sticky tongue and keen sense of smell. Sometimes a mother pangolin rolls her sleeping, curled-up baby to safety to protect it. But for pangopups, a mother's protective instincts can go only so far: These solitary, secretive animals are intensely hunted for bushmeat and for use in traditional medicine. Among the four African species, white-bellied pangolins are most at risk.

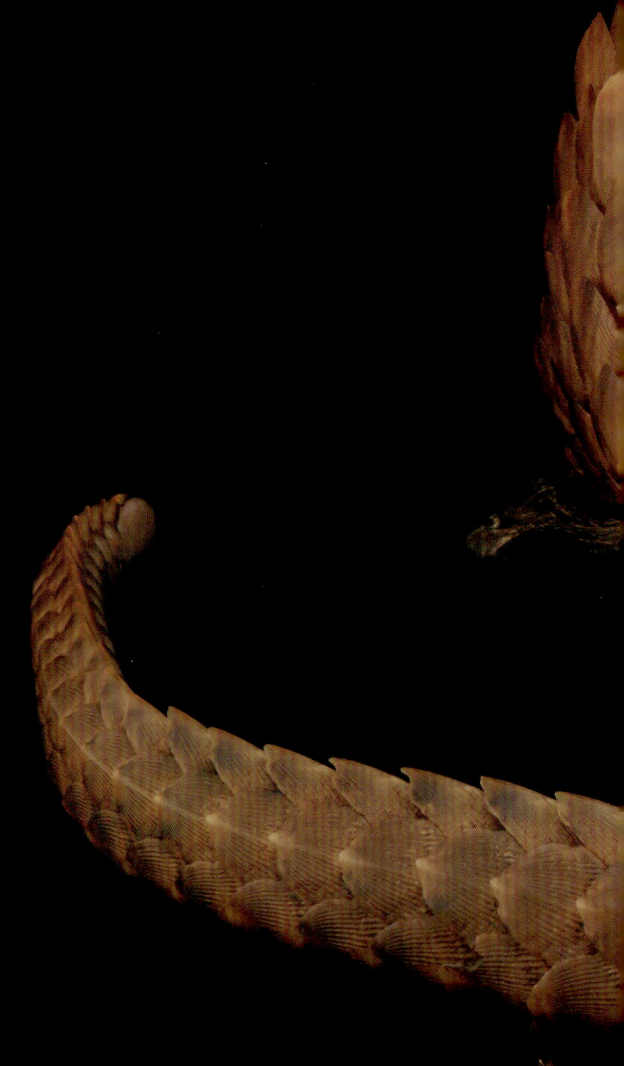

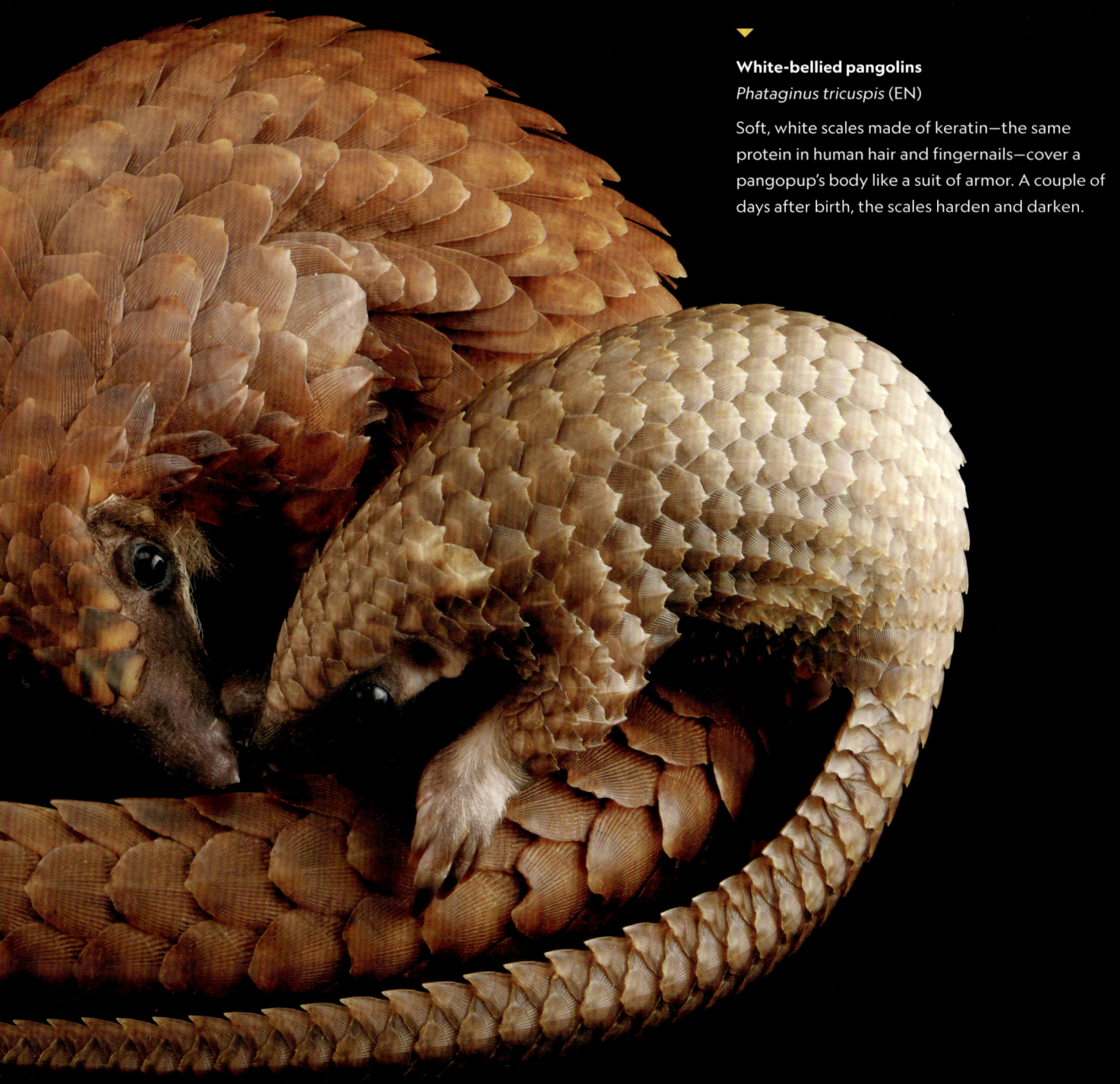

White-bellied pangolins
Phataginus tricuspis (EN)

Soft, white scales made of keratin—the same protein in human hair and fingernails—cover a pangopup's body like a suit of armor. A couple of days after birth, the scales harden and darken.

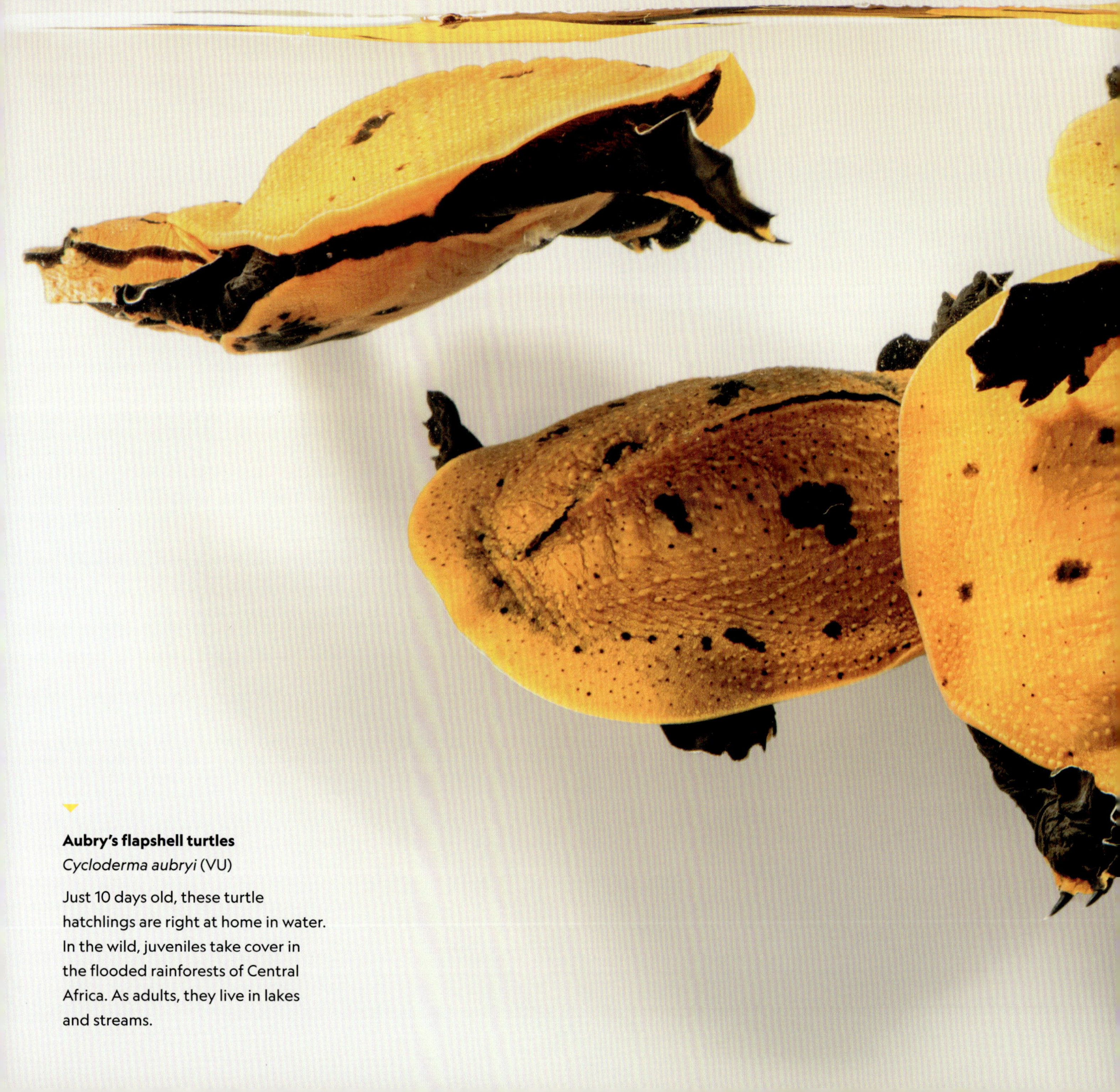

Aubry's flapshell turtles

Cycloderma aubryi (VU)

Just 10 days old, these turtle hatchlings are right at home in water. In the wild, juveniles take cover in the flooded rainforests of Central Africa. As adults, they live in lakes and streams.

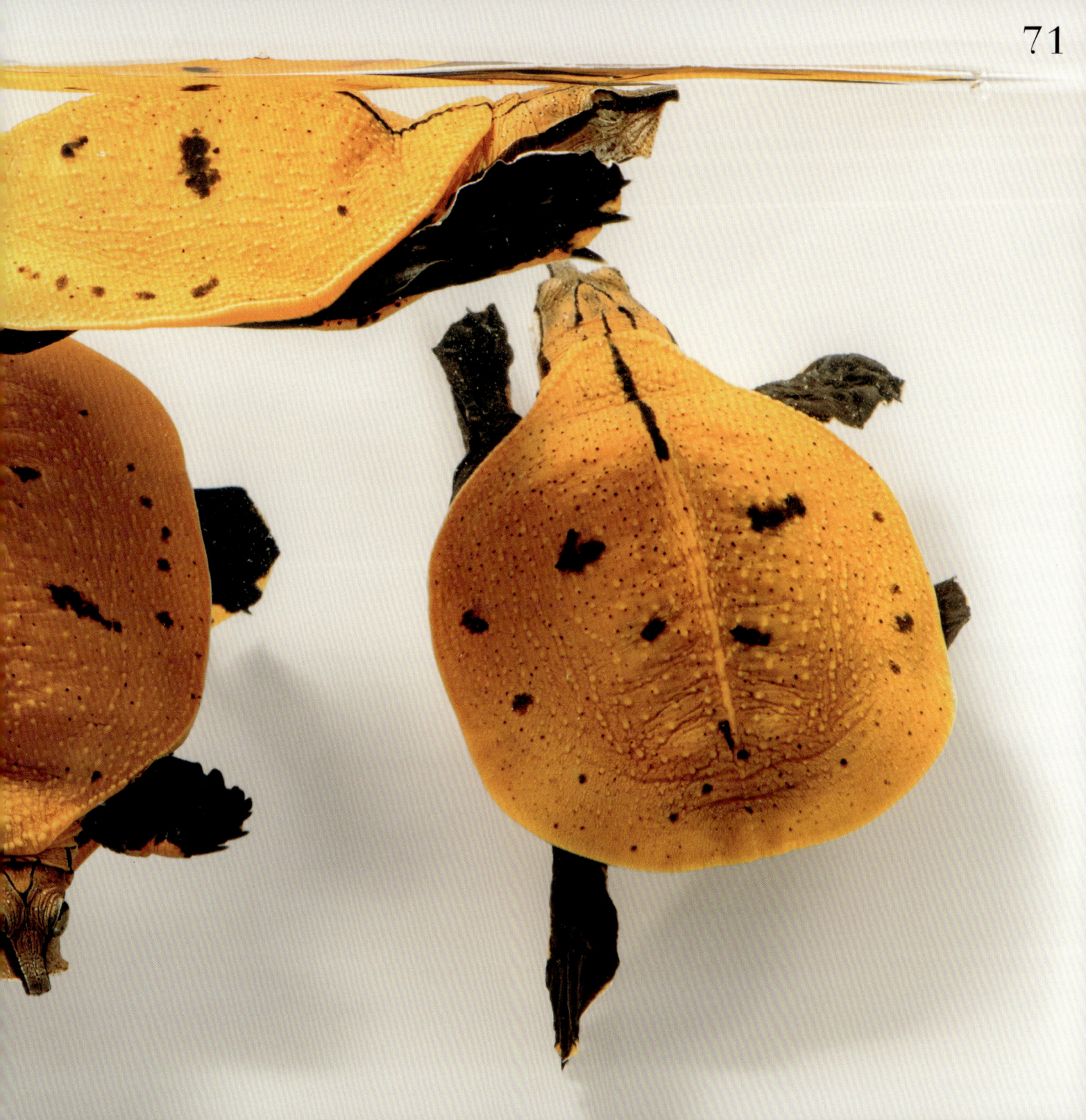

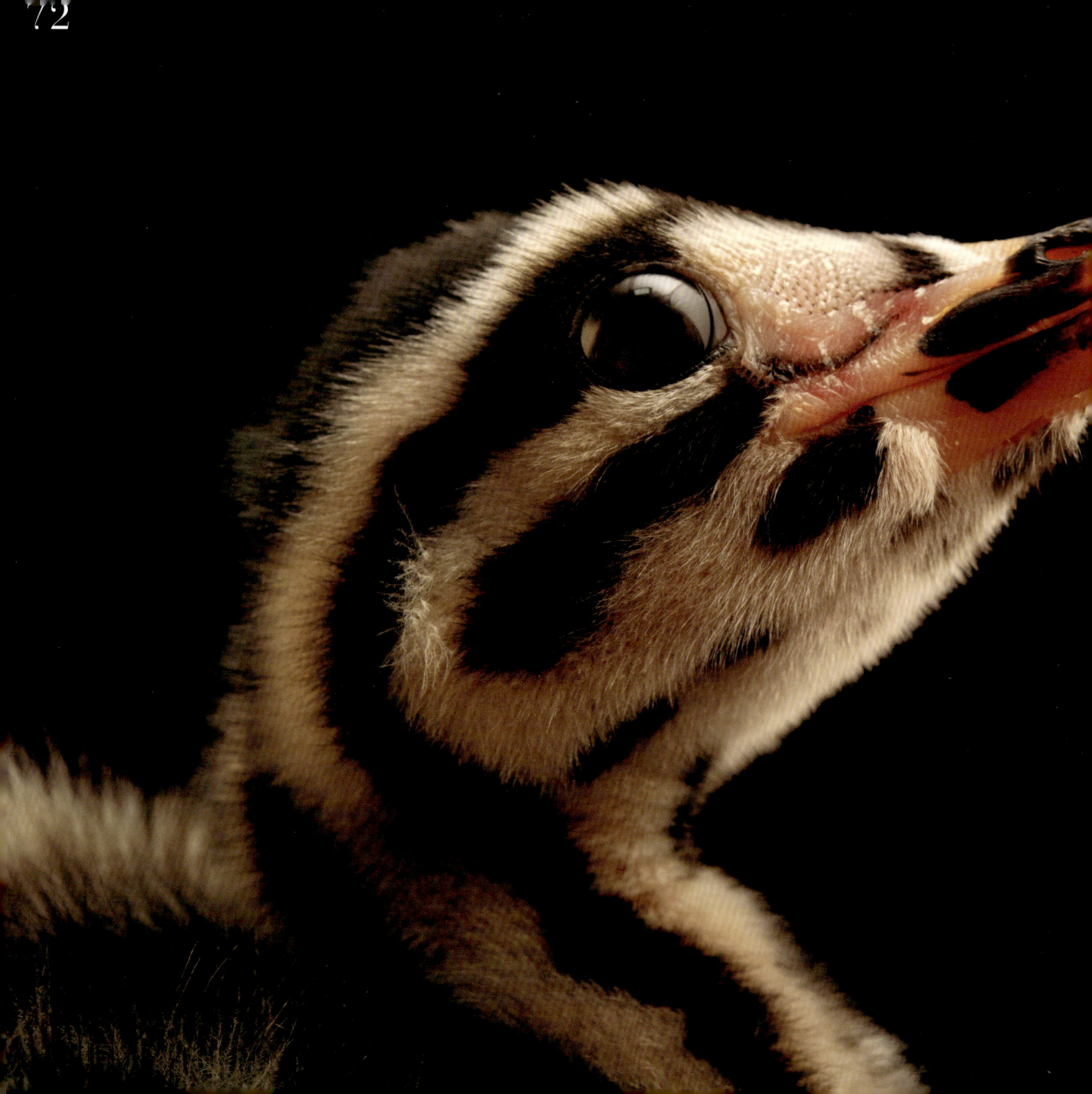

BEHIND THE LENS

WAKE UP, SLEEPYHEAD

ANIMAL: Red-necked grebe, *Podiceps grisegena holbollii* (LC)

LOCATION: Wildlife Rehabilitation Center of Minnesota, Roseville, Minnesota

"Baby animals don't have much of an attention span, so we really have to hurry up when we're taking their picture. We also have to keep them warm, so we always work in a temperature-controlled room. First, we talk to the keepers and ask them, 'How long would you be comfortable having this baby on this background?' Baby songbirds are the sleepiest. They don't look at us for long. And ducklings are famous for imprinting on people. It's very hard to keep them from coming right up to us during the shoot. They want to cuddle up. It's always a bit of a race. I have a helper place the ducklings from atop the shooting tent; we say, 'One, two, three!' and then the ducklings race toward me as I quickly snap a picture."

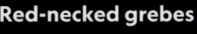

Red-necked grebes

Podiceps grisegena holbollii (LC)

Not long after hatching, a grebe chick will snuggle beneath the soft feathers of a parent's back to warm up. Then the family paddles away from its lakeshore nest into open water.

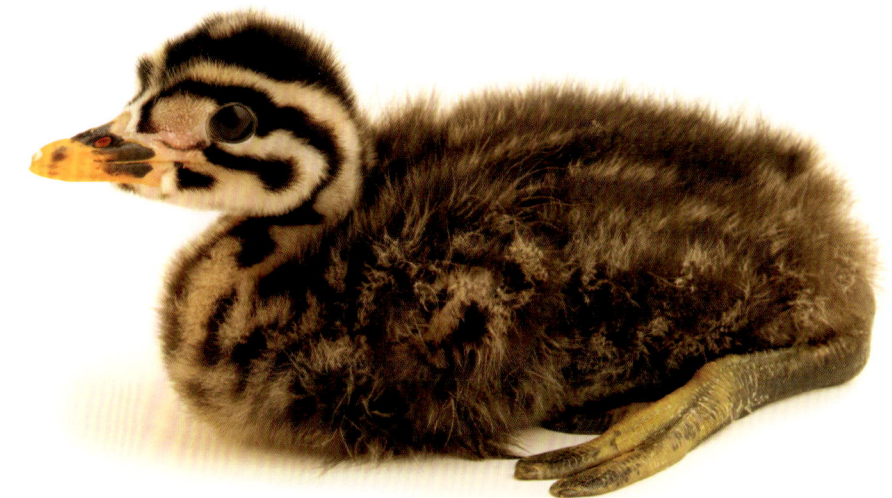

▼

Peruvian spider monkey
Ateles chamek (EN)

This little acrobat's long fingers and prehensile tail help it swing in search of fruit among the treetops of Bolivia, Brazil, and Peru. Youngsters stay close to their moms until they're 15 months old.

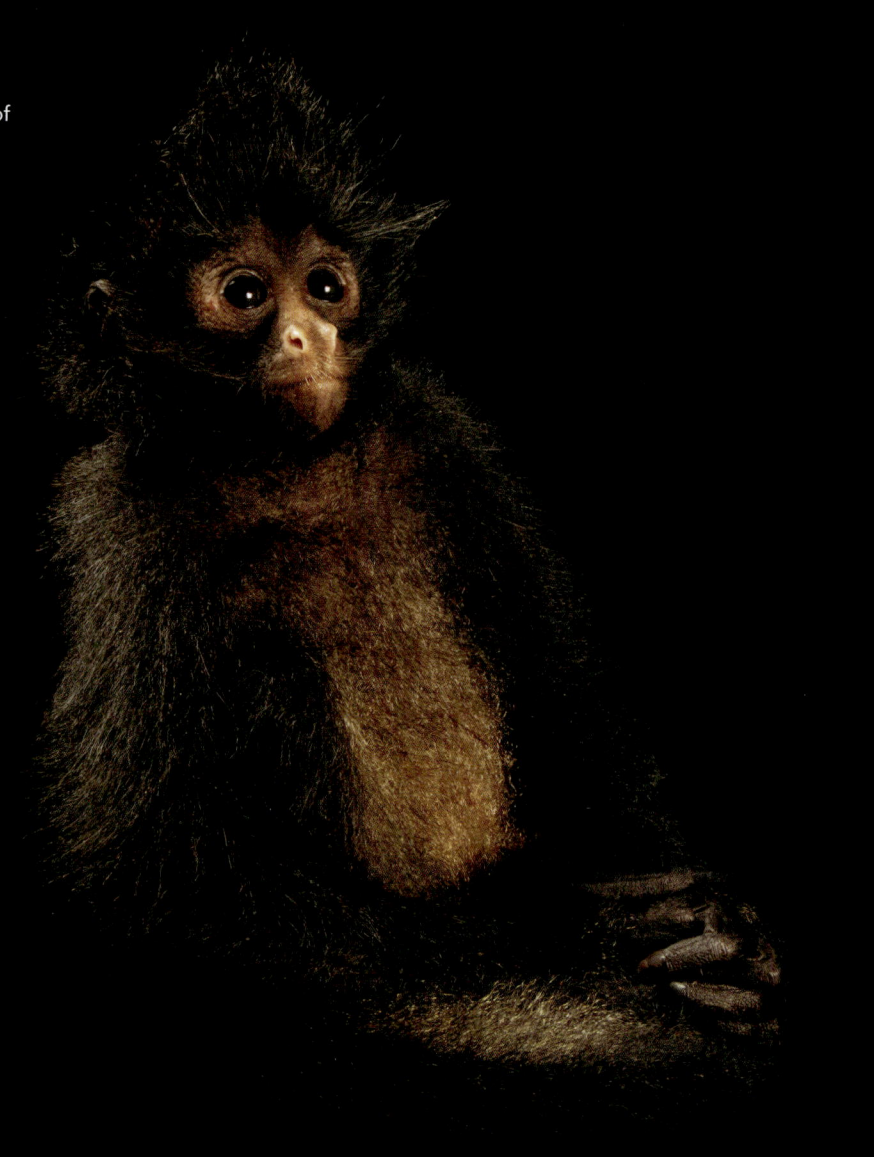

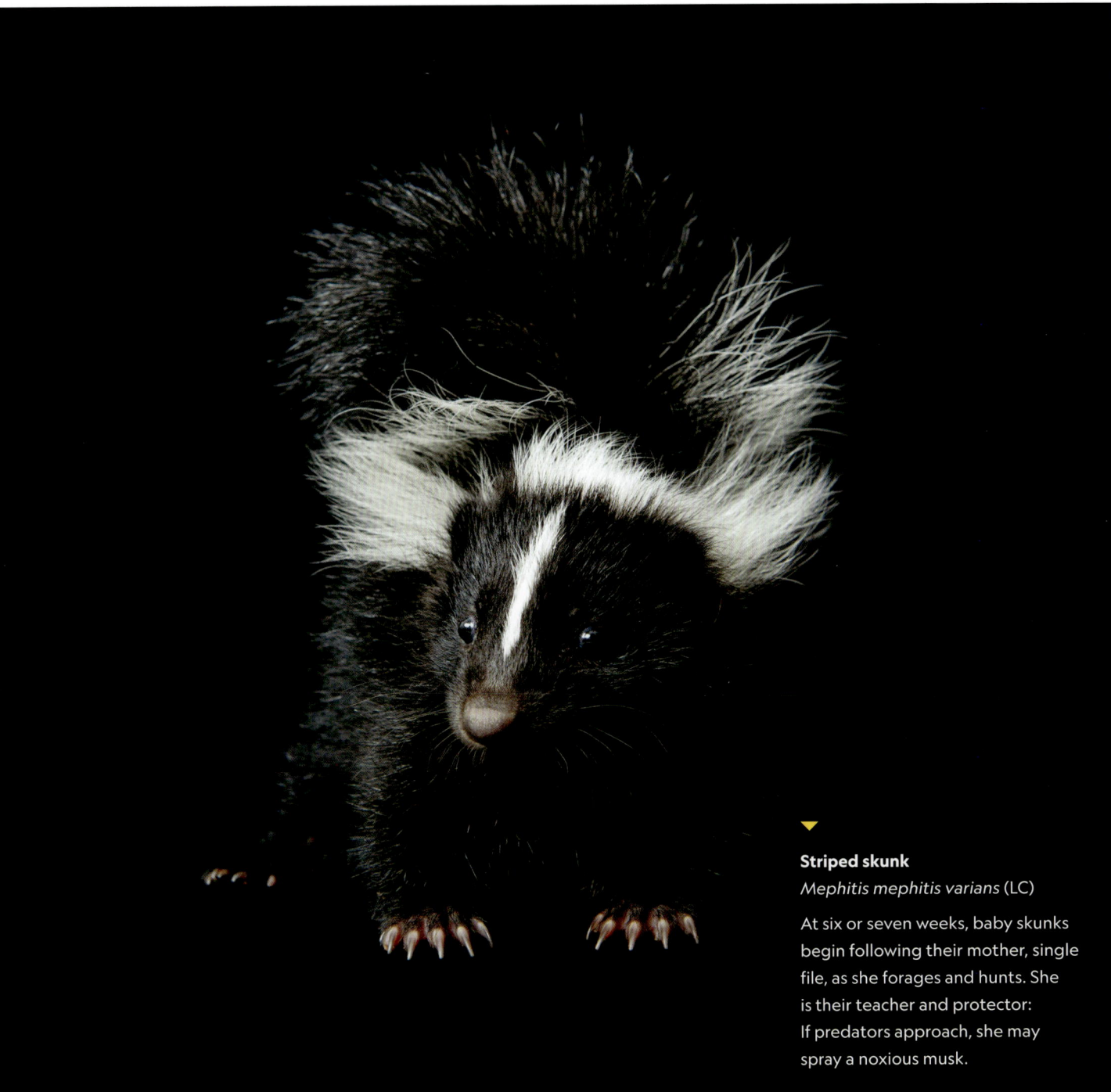

Striped skunk

Mephitis mephitis varians (LC)

At six or seven weeks, baby skunks begin following their mother, single file, as she forages and hunts. She is their teacher and protector: If predators approach, she may spray a noxious musk.

READY, SET, JUMP!

Leap of Faith

B

BUFFLEHEADS ARE SEA DUCKS, but they nest and raise their young in the boreal forests of North America. These ducklings begin their lives high up in a cozy tree nest, which their mother has lined with her soft down feathers. They can walk, jump, and swim right after they hatch.

Ducklings soon put their precocious skills to the test: About a day after hatching, the little ducks teeter on the edge of their nest and leap—sometimes as far as 46 feet (14 m)—to the ground. Their mom is the last to leave the nest; once they all reunite on the ground, she escorts her ducklings to a safe pond nearby. This short walk from the nesting tree to the pond is one of the few times buffleheads will walk on land; otherwise, they spend most of their lives in water, either the ponds where they grow up or the coastal waters where they spend winters.

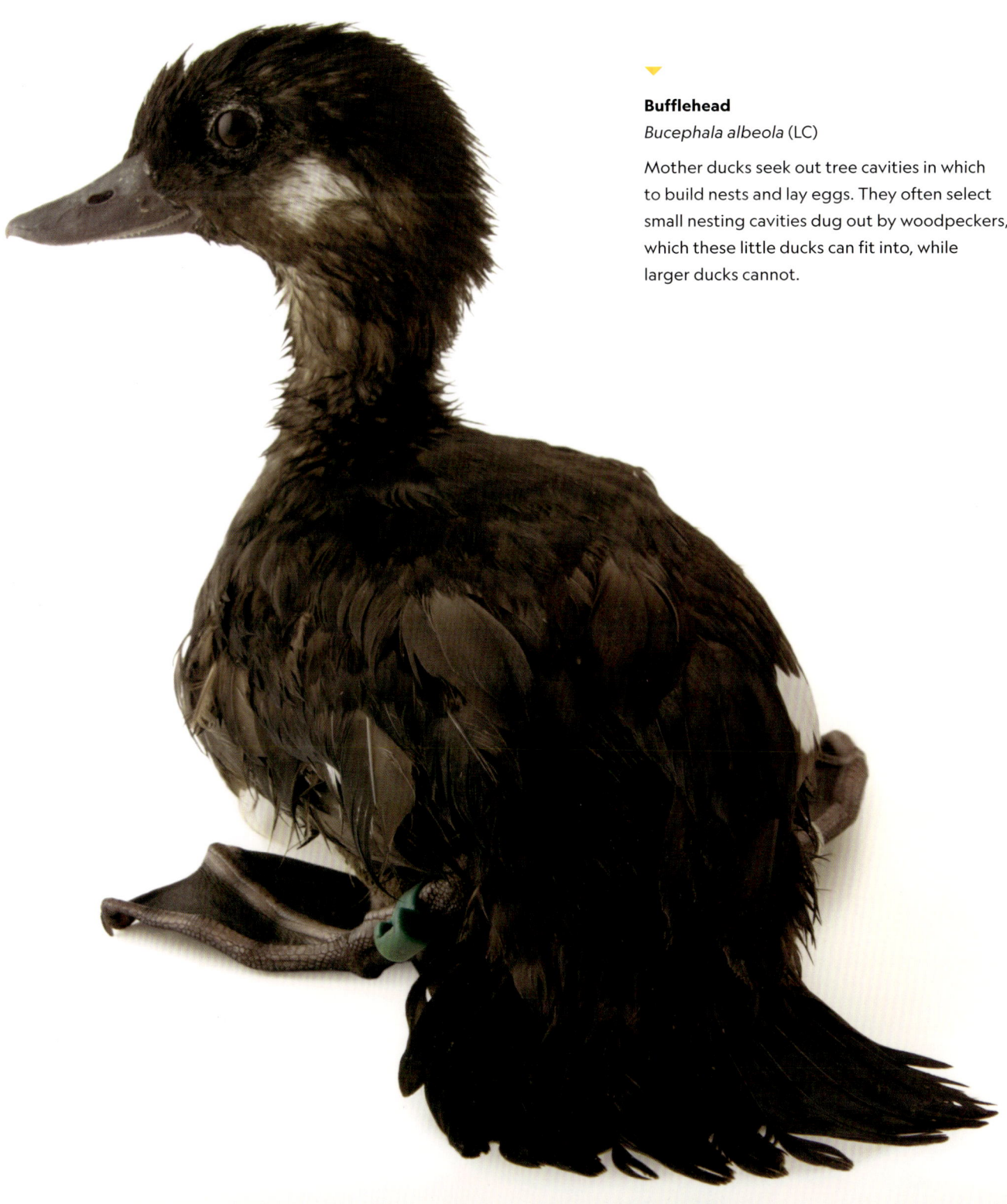

Bufflehead

Bucephala albeola (LC)

Mother ducks seek out tree cavities in which to build nests and lay eggs. They often select small nesting cavities dug out by woodpeckers, which these little ducks can fit into, while larger ducks cannot.

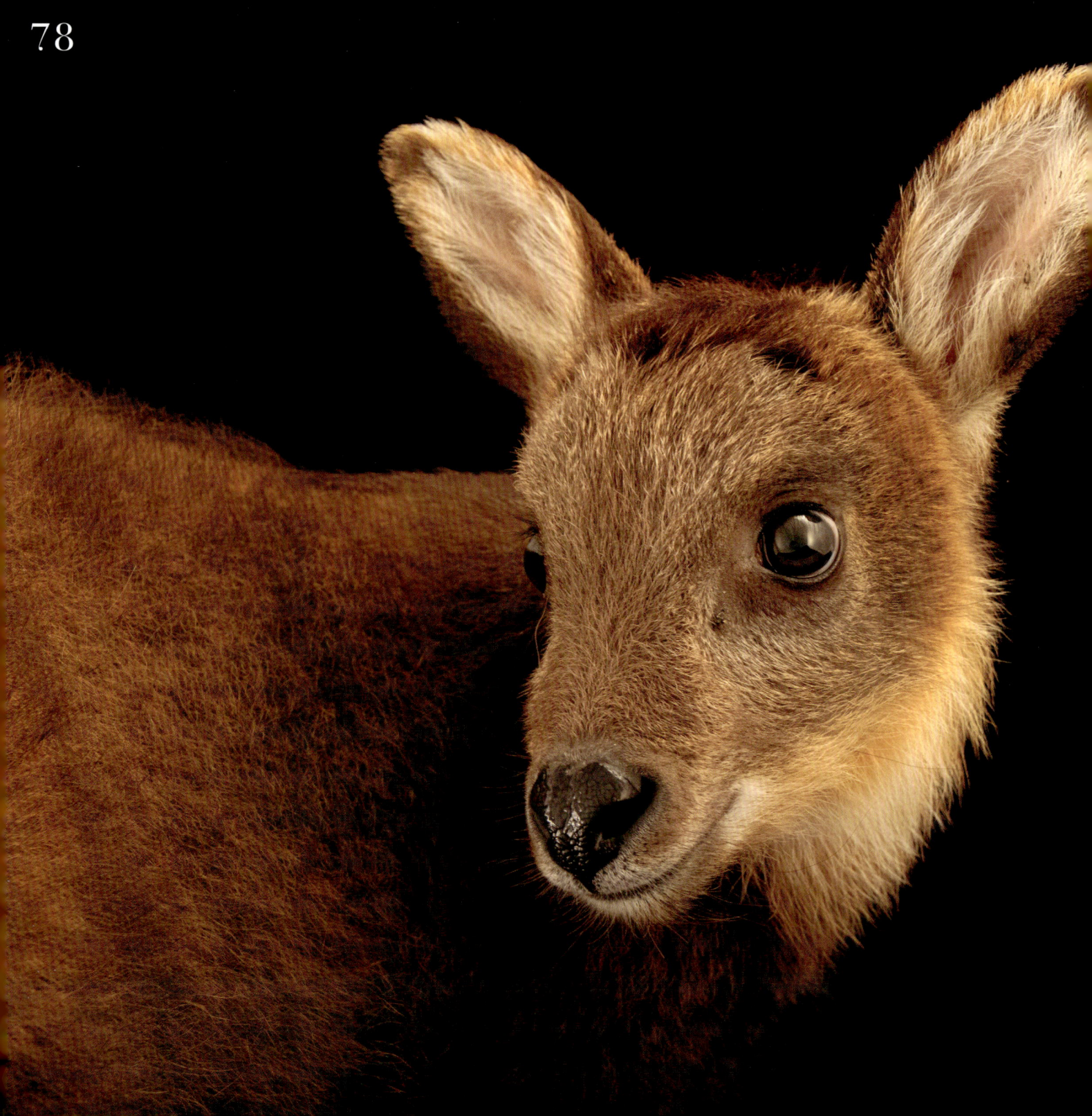

> 66
>
> "This picture speaks to the nature
> of all wild animals. Even though it's
> a baby, it's alert, it's smart, it's curious—
> for if it is not, it won't last a day
> once it leaves its mother."

▼

Formosan serow

Capricornis swinhoei (LC)

In the wild, a month-old baby like this begins
following its mother around immediately
after birth. The rugged mountains and rocky
areas are no problem; this little one is
a born climber.

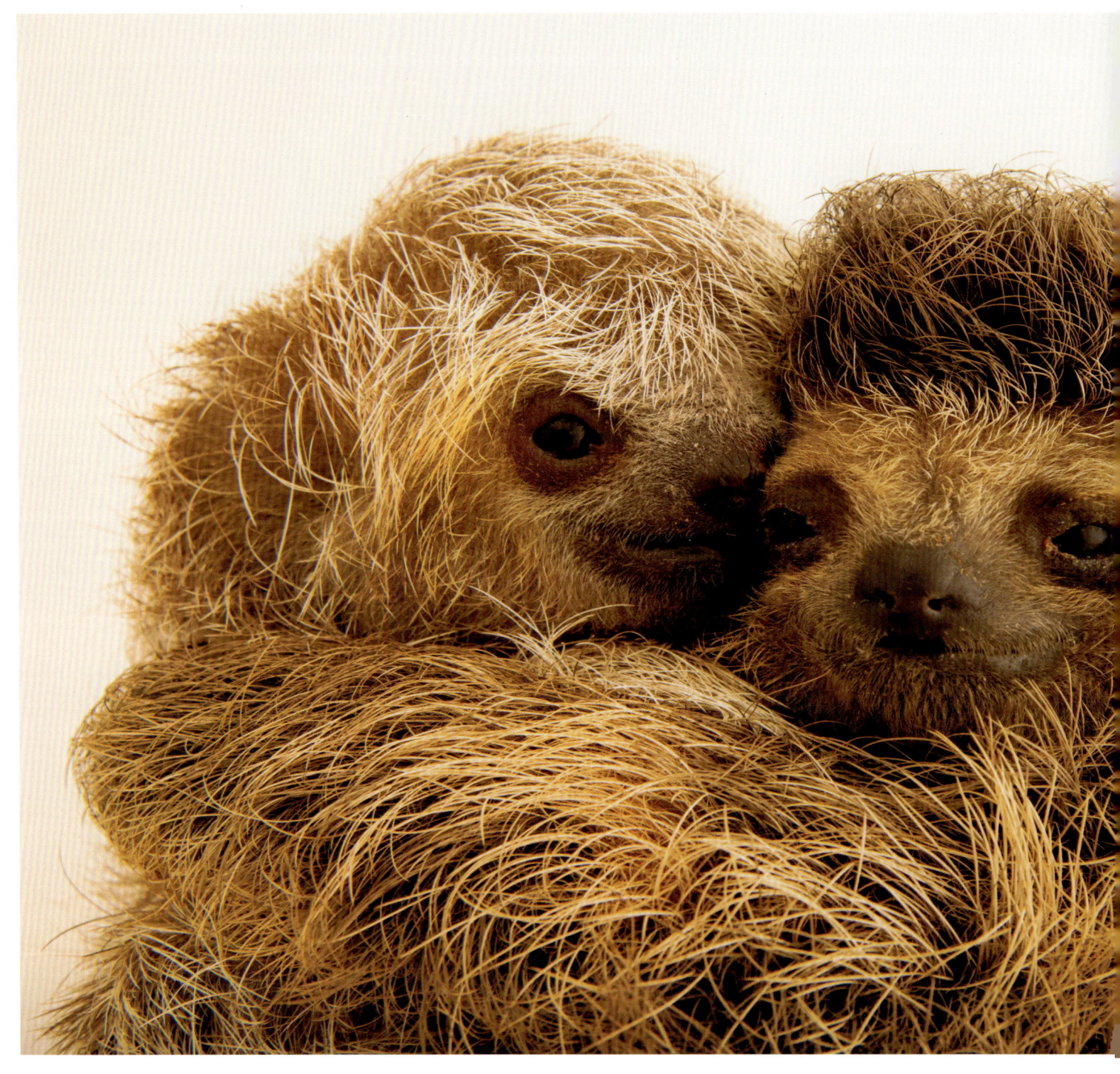

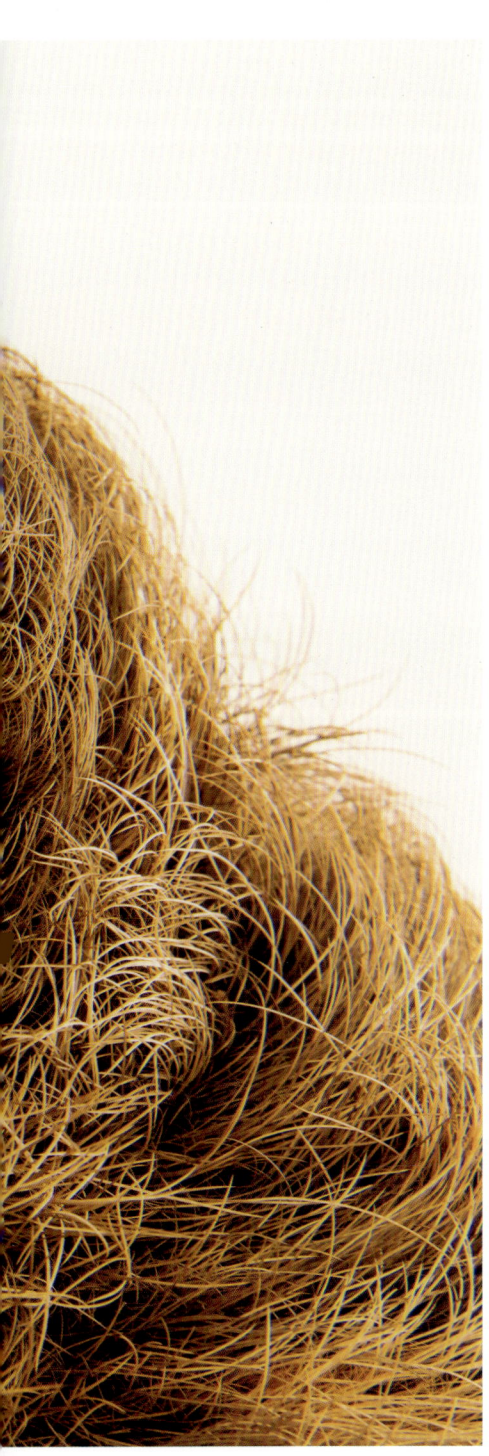

Brown-throated sloths

Bradypus variegatus (LC)

When a female gives birth, often in a tree, other sloths make sure the little one doesn't fall while they clean the mother and baby. Mothers help their infants develop motor skills and teach them how to explore.

SAVING THE SPECIES

INTO THE WILD

Every year, human caretakers at places like Costa Rica's Toucan Rescue Ranch and the Sloth Institute rescue orphaned sloths from the rapidly changing forests nearby. The little sloths have lost their homes to deforestation or their parents to electrocution, as sloths living in fragmented habitats sometimes must scale electrical wires to reach food. Like other mammals with close maternal bonds, baby sloths require long periods of nurturing and teaching to be ready for life on their own. But what happens when an orphaned sloth has no mother to care for it?

That's where "sloth school" comes in. At the Sloth Institute, wildlife rehabilitators have developed a multistep program for juvenile sloths to make sure they receive the care they need to safely return to their forest habitat. In sloth elementary school, the assignments are simple but vital: Eat, grow, and stay healthy. In middle school, young sloths learn climbing, agility, and social skills. By high school, they are ready to forage and build their confidence. And what about university? That's when sloths, who sleep up to 20 hours a day, have to master an essential skill: how to pick a safe spot to sleep. If an orphaned sloth has reached a healthy weight and has mastered these important life skills—a process that can sometimes take up to two years—it is released back into the wild.

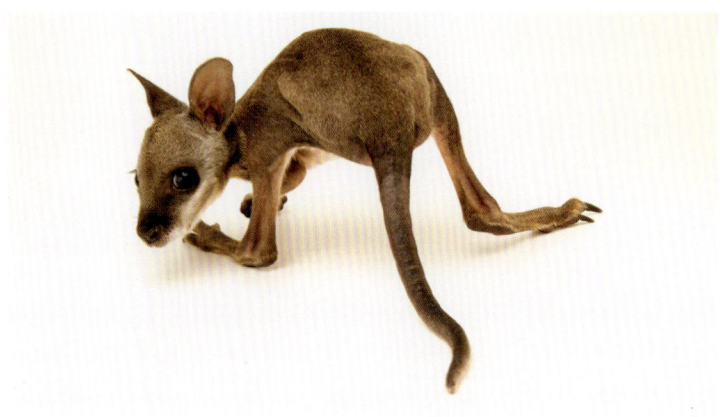

Tammar wallaby *Notamacropus eugenii* (LC)

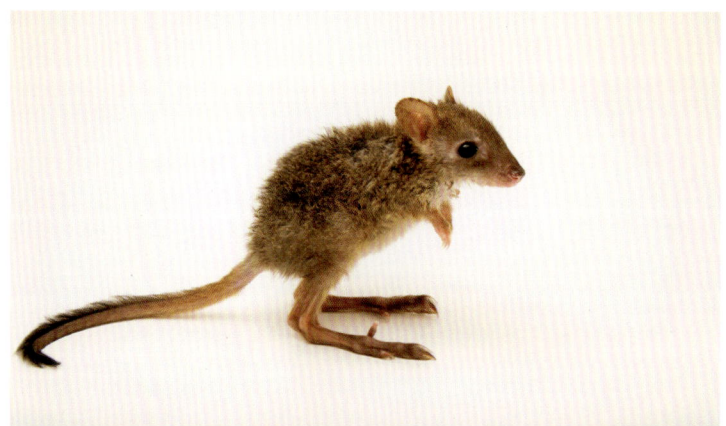

Brush-tailed bettong *Bettongia penicillata* (CR)

LIFE IN THE POUCH

▼

Australia is home for all these marsupials—mammals whose young are born so early, they can survive only by nursing and growing while tucked in their mother's pouch. Time spent in the pouch varies: This 20-week-old bettong joey (center right) has been out of the pouch for about three months; in the wild, an eight-month-old swamp wallaby like this one (bottom right) could still be in its mother's pouch.

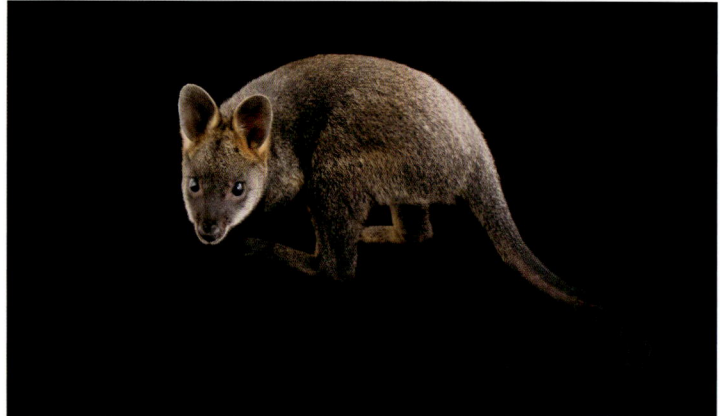

Swamp wallaby *Wallabia bicolor* (LC)

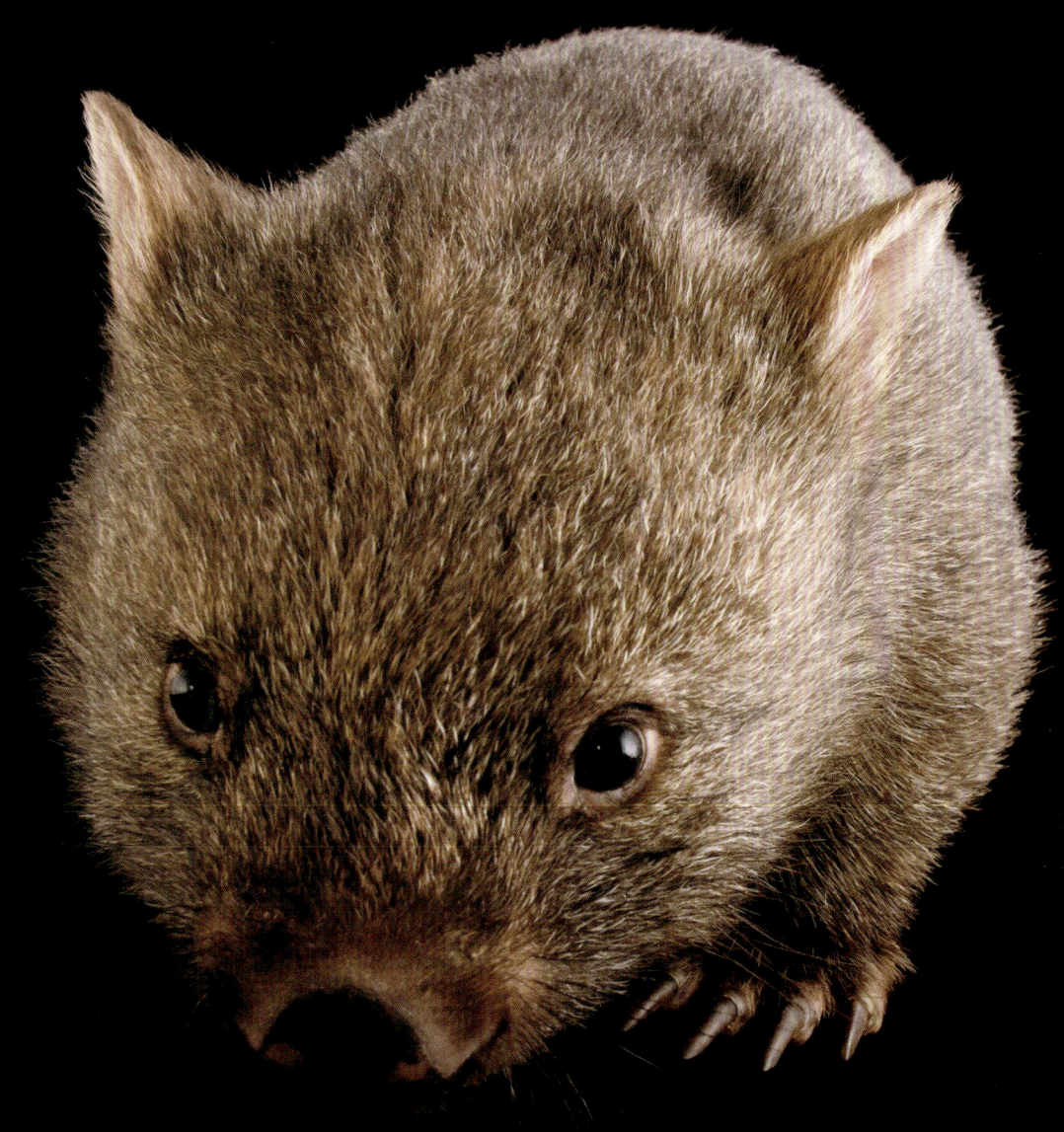

Common wombat

Vombatus ursinus tasmaniensis (LC)

In the wild, wombats are born the size of a bean and stay in their mother's pouch for six months; they remain dependent on her until they're 18 to 20 months old. This youngster is 10 months.

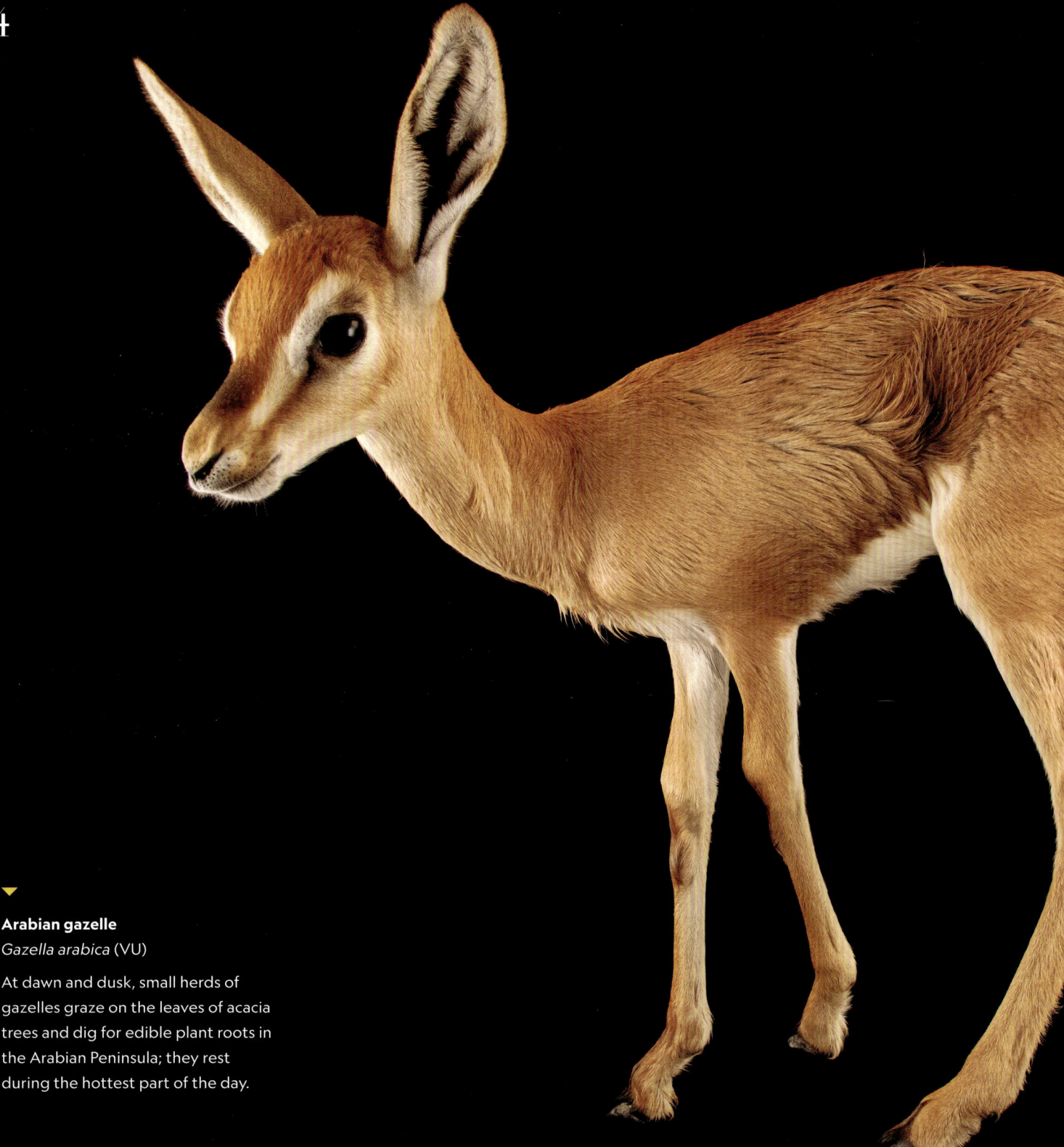

Arabian gazelle
Gazella arabica (VU)

At dawn and dusk, small herds of gazelles graze on the leaves of acacia trees and dig for edible plant roots in the Arabian Peninsula; they rest during the hottest part of the day.

QUICK LEARNER
Born to Run

F

FROM THE VERY first minutes of their lives, gazelles' survival depends on their ability to stand up, walk, run, and follow the herd. Like many hoofed mammals around the world—horses, antelope, caribou—this sure-footed resident of the deserts, hills, and coastal plains of the Arabian Peninsula is alert from the start. Females give birth to a single fawn at any time of year, although spring and fall are most common. Fawns nurse from their mother for six months.

After that, young male gazelles leave their mothers and set off to join a herd of bachelor gazelles. A female may live her entire life with her mother in a small group along with other female gazelles and their young.

Ranges stone geckos
Diplodactylus furcosus (LC)

Females of this species, native to the interior of Australia, usually lay two eggs at a time, but they can have multiple clutches in a year. Pictured here is an adult male with a one-month-old juvenile.

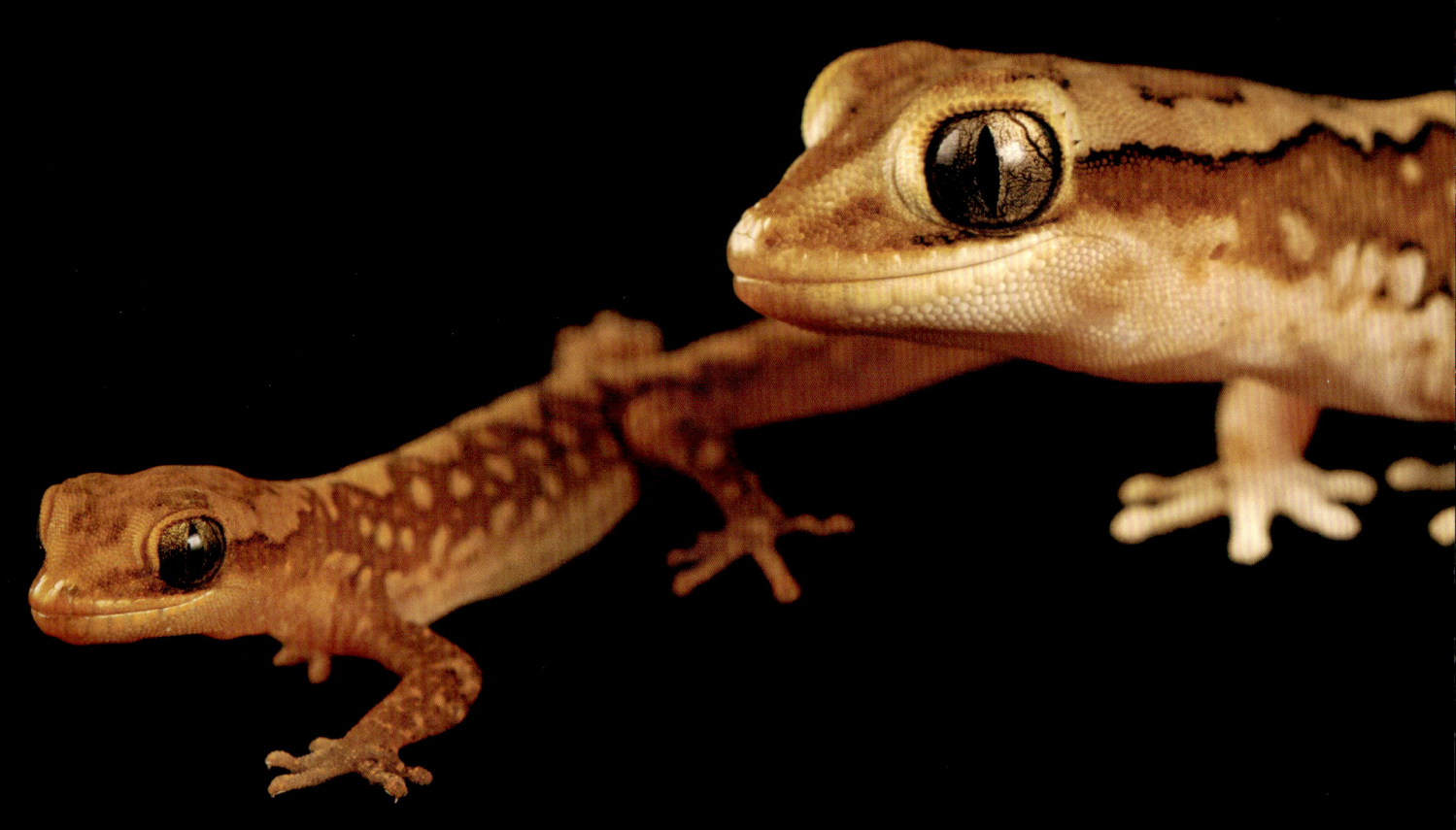

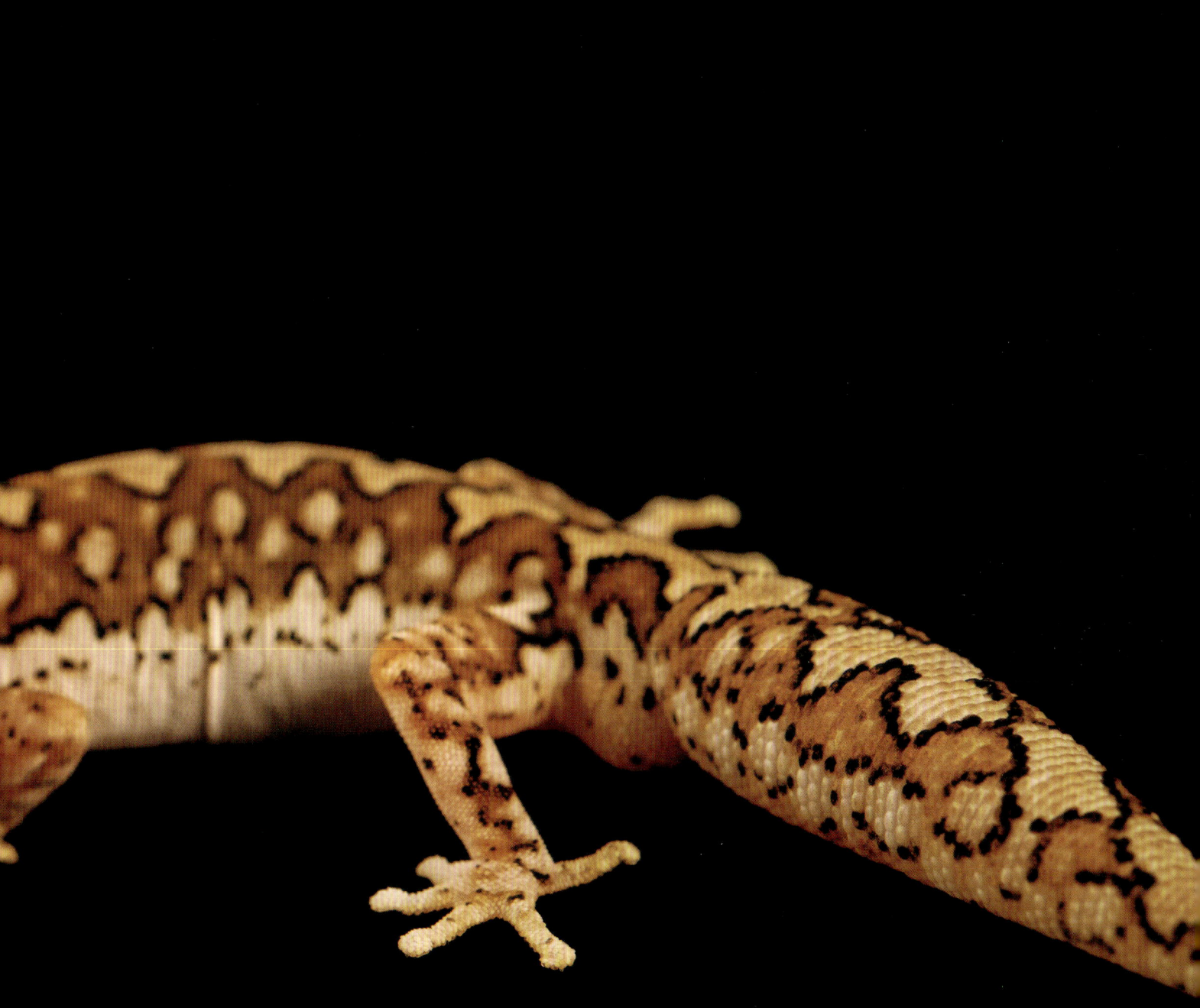

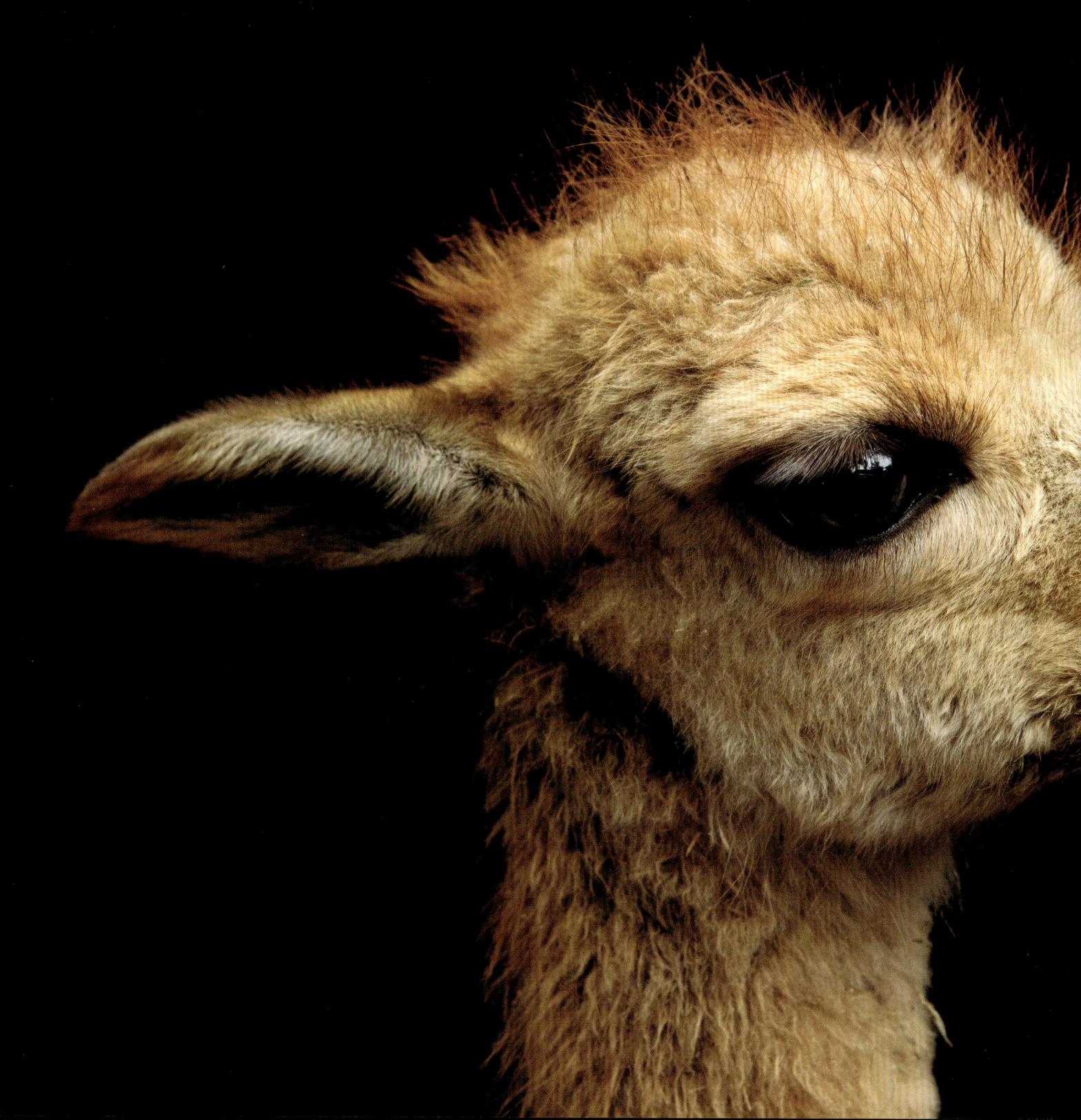

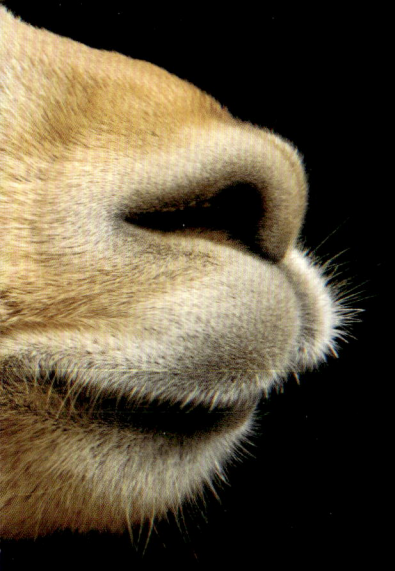

3 | YOUTH

Growing Up Is Hard to Do

It's a tough world out there for youngsters.
Rugged terrain, predators, extreme temperatures, disease—danger lurks
all around. That's why nature has provided youthful superpowers.

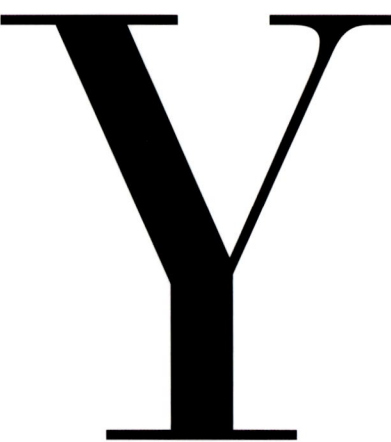

YOUNG ANIMALS ARE masters of disguise, savvy tricksters, and nimble acrobats. They know all the good places to hide, and their bodies are brilliantly designed to keep them safe. Sometimes they get assistance from the adults in their lives, who offer protection or food. But young animals often have to rely on their smarts, instincts, and physical characteristics to survive.

Caterpillars fend off would-be assailants with stripes and spikes that signal their toxicity. Tree frogs give much the same warning with their vivid colors. Pangopups curl into a tight ball; lemon shark pups take cover among tangled mangrove roots beyond the reach of larger predators; and dappled fawns wait in the undergrowth for their mothers, knowing instinctively to hold absolutely still.

The physical changes young animals experience as they journey toward adulthood can be subtle or dramatic. As they trade their nurseries for the broader world—moving from ground to tree, or vice versa—their colors and markings sometimes change too.

A baby tapir will lose its spots and stripes, maturing to look like an enormous black-and-white cookie. Young zebra sharks trade their stripes for leopard-like spots. A green tree python shifts from red or yellow to green. Some freshwater fish, such as the redspot darter, drab as juveniles, announce their maturity with markings of vibrant red and blue.

When not dodging predators or watching their footing, young animals are learning what it takes to survive. Their behaviors change. Tiger and cheetah cubs, for instance, learn important hunting behaviors through fierce play with their siblings, but as they mature, that play becomes the real deal. By following their mom day after day and watching how she stalks and captures her next meal, they learn how to do it themselves.

For all their experimentation with adult behaviors, though, there are plenty of awkward moments. These small fry with their gangly legs, oversize paws, big ears, and wayward feathers aren't quite grown. Beaver kits, born with thick fur and able to swim right away, nonetheless stick around their family's den for two or three years. Romping with their siblings, slapping their tails, sleeping in a warm den: Life is good for busy young beavers. But by their first autumn, the youngsters get a glimpse of what's ahead. With their parents and older siblings, the kits dig canals and fetch sticks to store in an underwater feed pile. Being cute, it turns out, isn't quite enough to get a pass on doing chores.

▼

Prairie gray fox *Urocyon cinereoargenteus ocythous* (LC)
Previous pages: **Peruvian vicuña** *Vicugna vicugna* (LC)

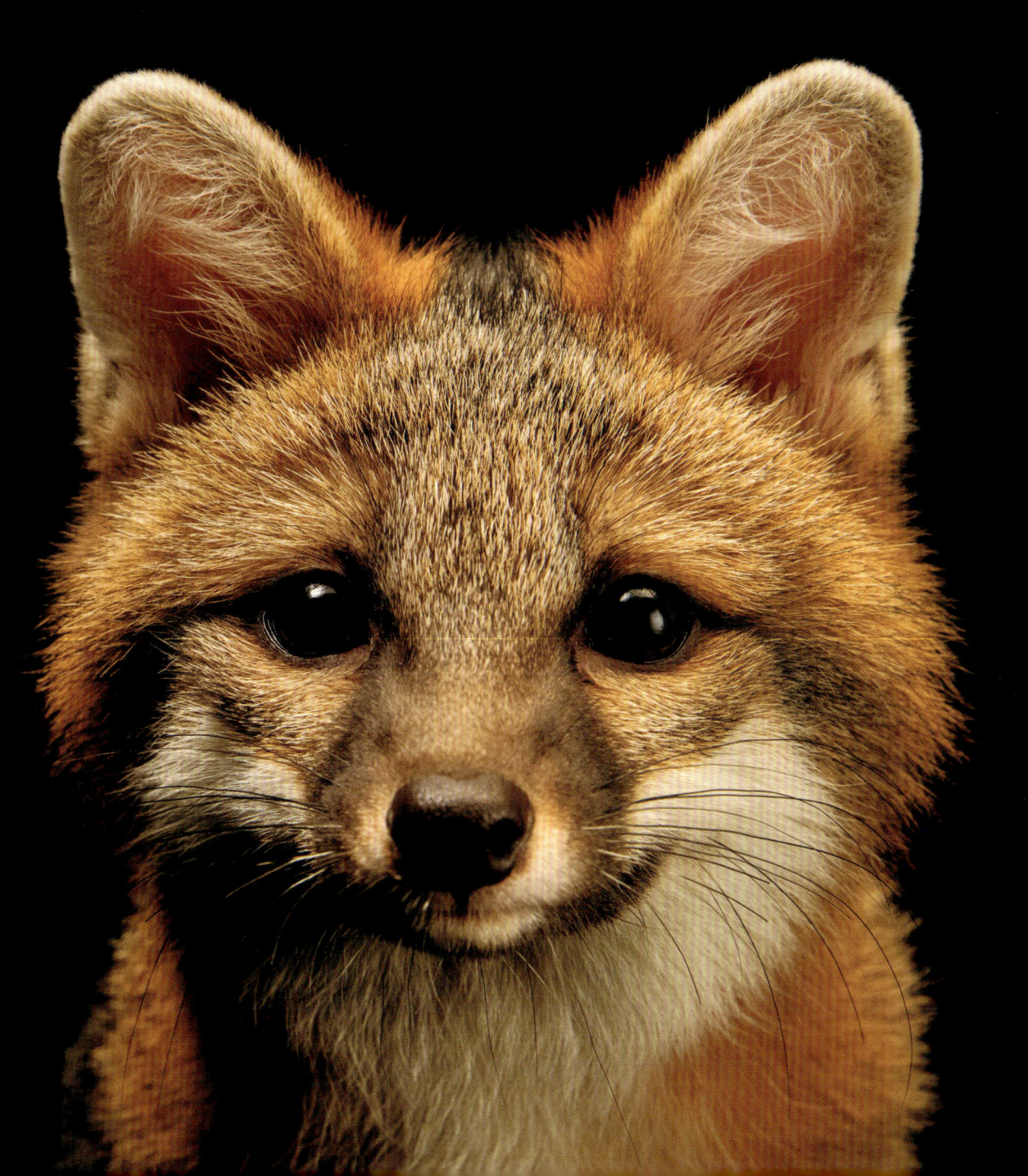

Don't Let Go
Family Ties

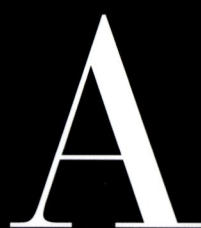

A

AS SOON AS THEY are born, baby northern white-cheeked gibbons hold on to their mothers—and they don't let go for two more years. They cling to her as she navigates the treetops in search of fruit and leaves to eat from the forests of Laos and Vietnam. Although these gibbons are born a creamy beige, by the time they are two years old—and often sooner—they turn black, with distinctive white cheek patches. (Females revert to beige when they're sexually mature; males remain black.)

Gibbons are physically independent when they're three years old, sexually mature around seven, and ready to leave their family by age eight. But some young gibbons are in no hurry to leave, and they choose to stay with their family until they're 10 years old.

Northern white-cheeked gibbons
Nomascus leucogenys (CR)

Infant gibbons like to groom one another, and playing with the youngsters is a priority for members of a gibbon family in their treetop home. The baby in this photo is a year old.

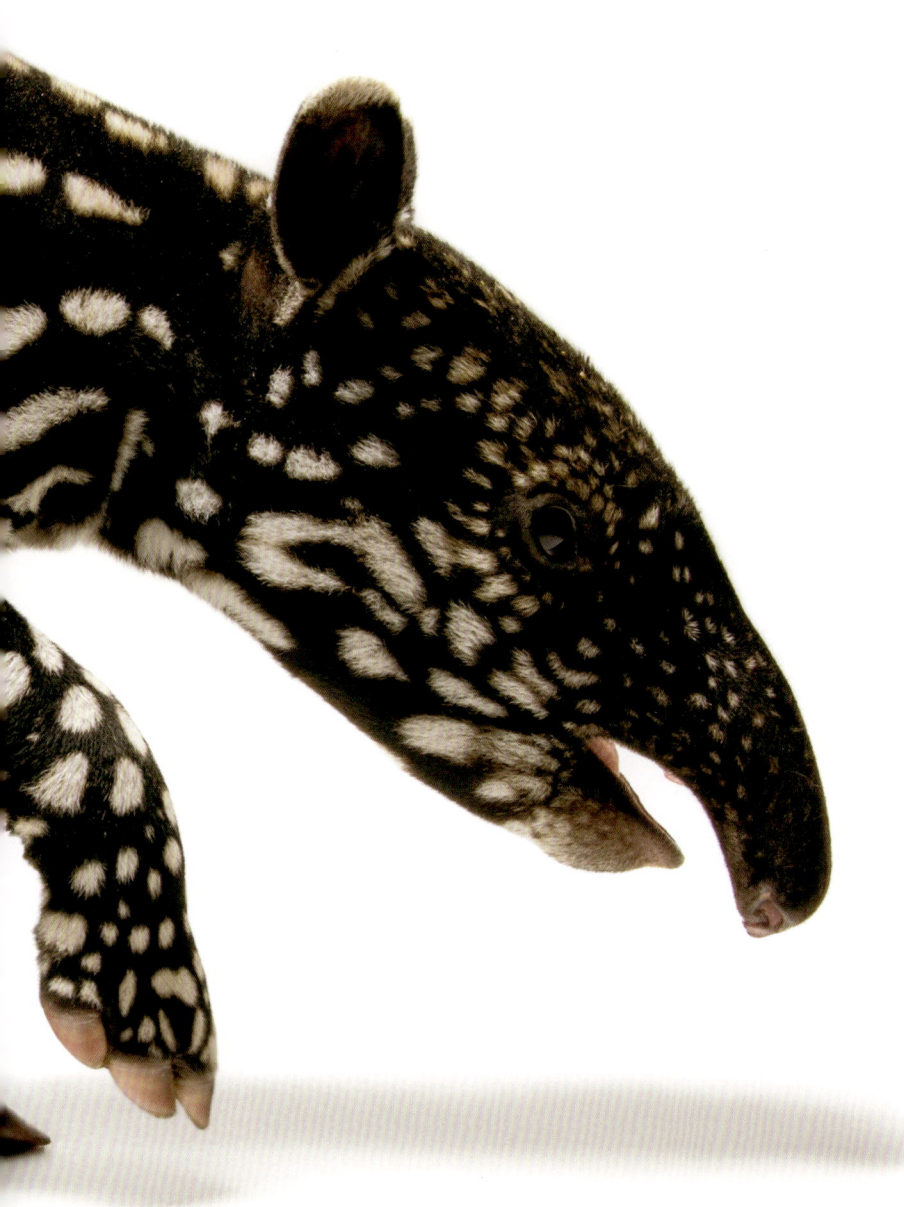

LOVE ME
Big Baby

Y

YOUNG MALAYAN TAPIRS have been perfecting their cuteness since ancient times. Indeed, scientists think these stocky animals have changed little for tens of millions of years.

Females have long pregnancies—between 11 and 13 months. After they give birth, their calves stay with them until the next calf is born, usually about two years later. Calves with stripes and spots are perfectly camouflaged for life in the sun-dappled forest, and they nurse from their mothers until they are six to eight months old. Tapirs are brave young souls, and rather than hide, they will rest, explore, and swim with their mothers. This calf radiates cuteness at just a few days old. But it will grow up, lose those markings, and eventually tip the scales at some 800 pounds (360 kg).

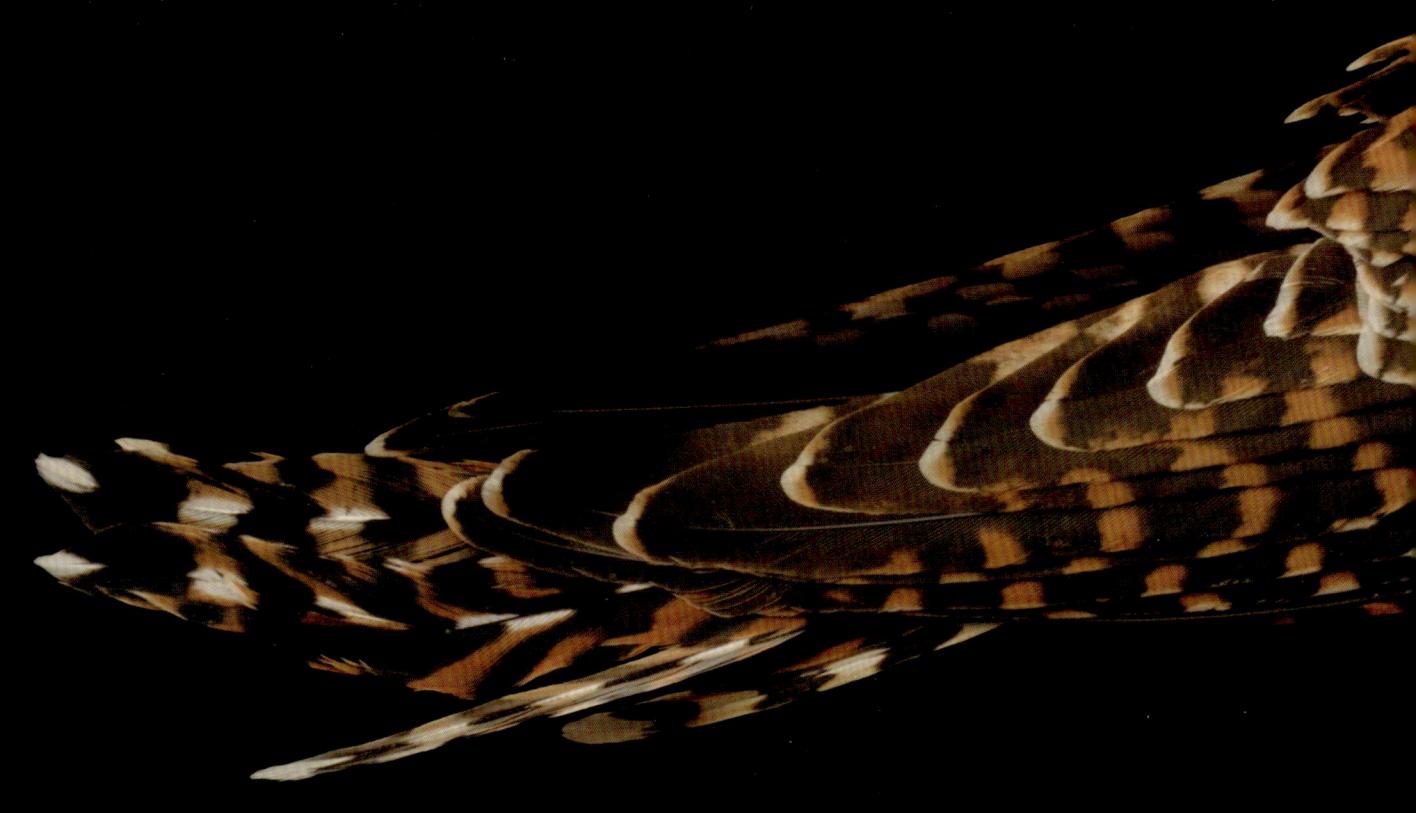

Common cuckoo

Cuculus canorus canorus (LC)

Female cuckoos forgo the hard work of chick rearing by laying eggs in the nests of other, smaller birds who care for the young, like this juvenile. Male chicks grow up to make the well-known cuckoo call.

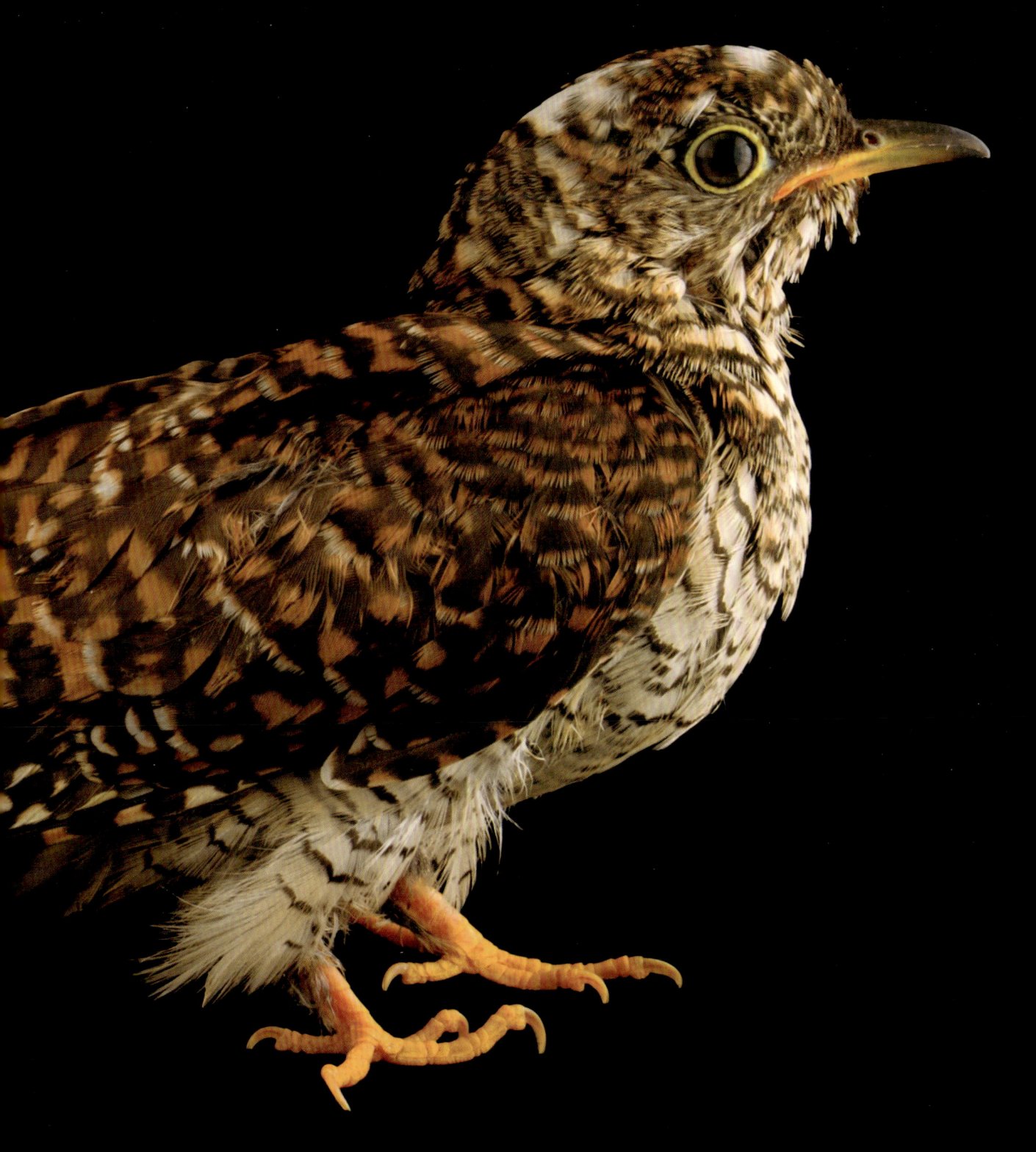

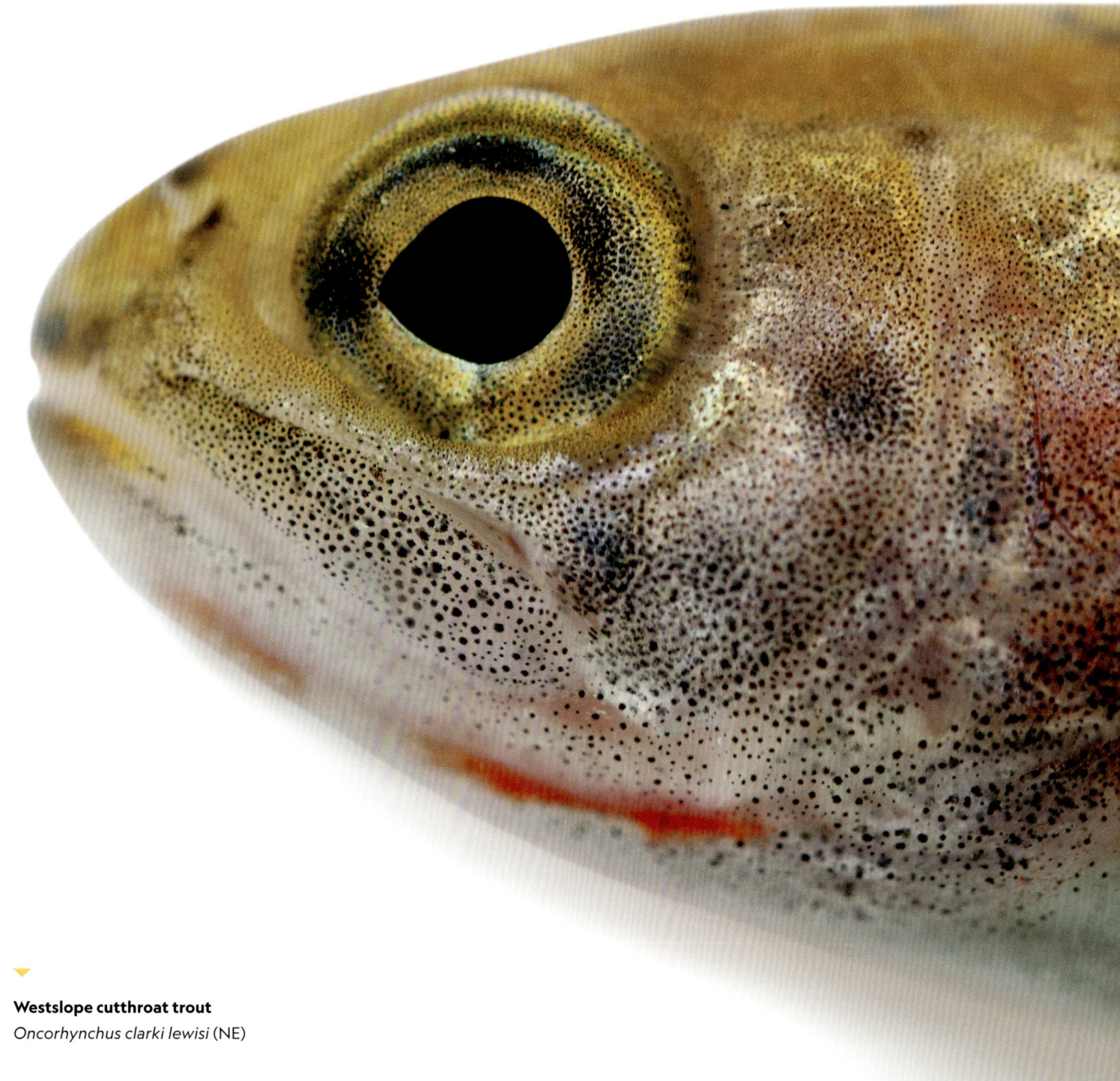

Westslope cutthroat trout
Oncorhynchus clarki lewisi (NE)

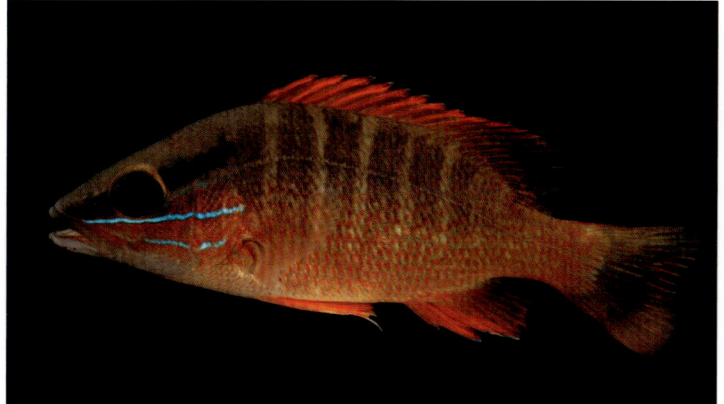

Delta smelt *Hypomesus transpacificus* (CR)

SWIM, BABY, SWIM

Life isn't easy for small fry, or newly hatched fish. Whether a fish species spawns amid a coral reef or a mountain stream, it needs a dash of luck for its fry to survive. This is especially true for California's tiny but critically endangered Delta smelt (top left), whose spawning habits are threatened by human-made hazards that affect water quality and flow.

Smallmouth bass *Micropterus dolomieu* (LC)

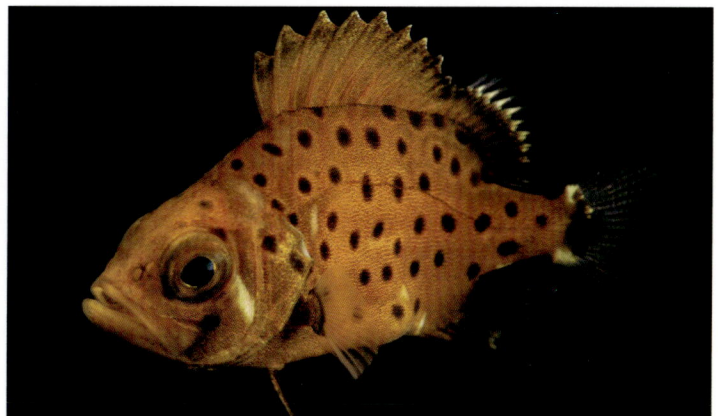

Giant sea bass *Stereolepis gigas* (CR)

Gorean snapper *Lutjanus goreensis* (DD)

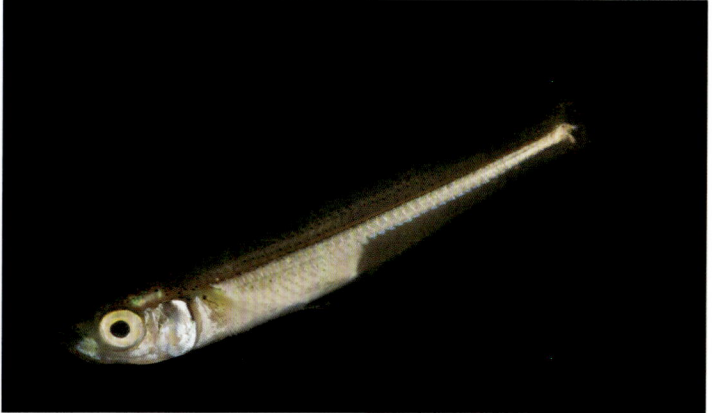

Atlantic silverside *Menidia menidia* (LC)

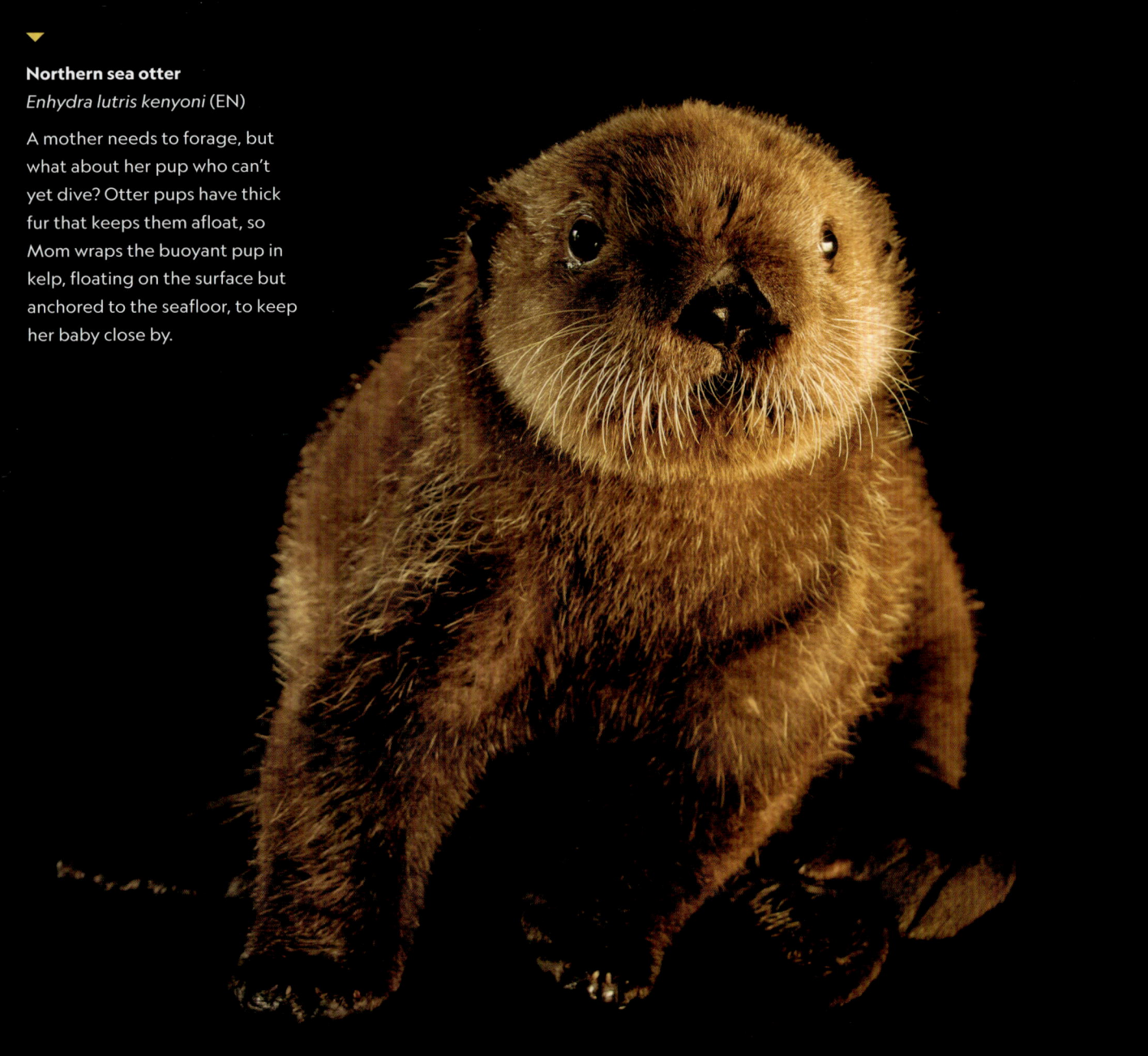

Northern sea otter

Enhydra lutris kenyoni (EN)

A mother needs to forage, but what about her pup who can't yet dive? Otter pups have thick fur that keeps them afloat, so Mom wraps the buoyant pup in kelp, floating on the surface but anchored to the seafloor, to keep her baby close by.

Eastern gray squirrel

Sciurus carolinensis carolinensis (LC)

Baby squirrels are born underdeveloped, helpless, and—aside from their whiskers—completely hairless. Under their mother's care, they grow quickly, turning into bushy-tailed backyard acrobats.

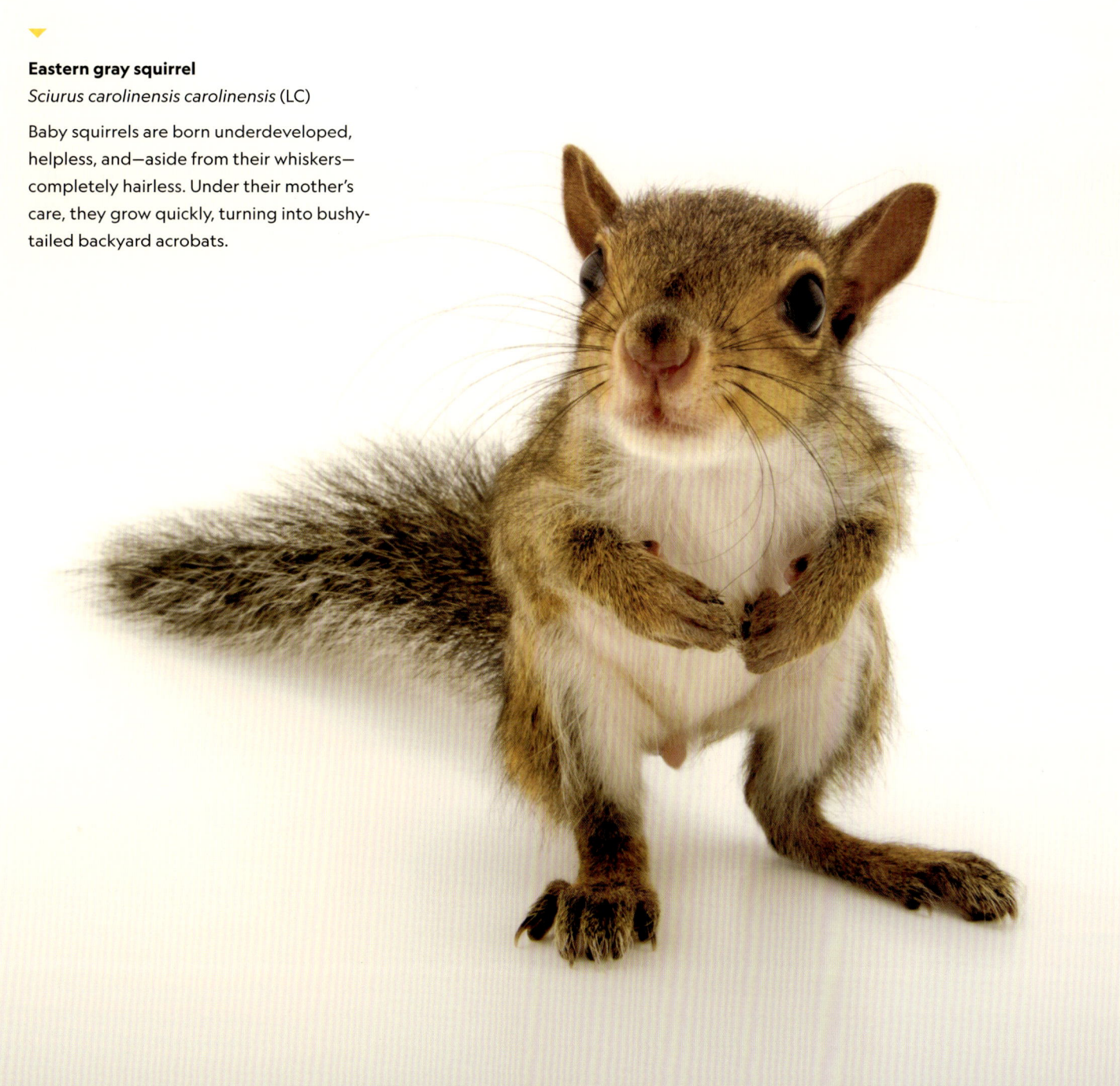

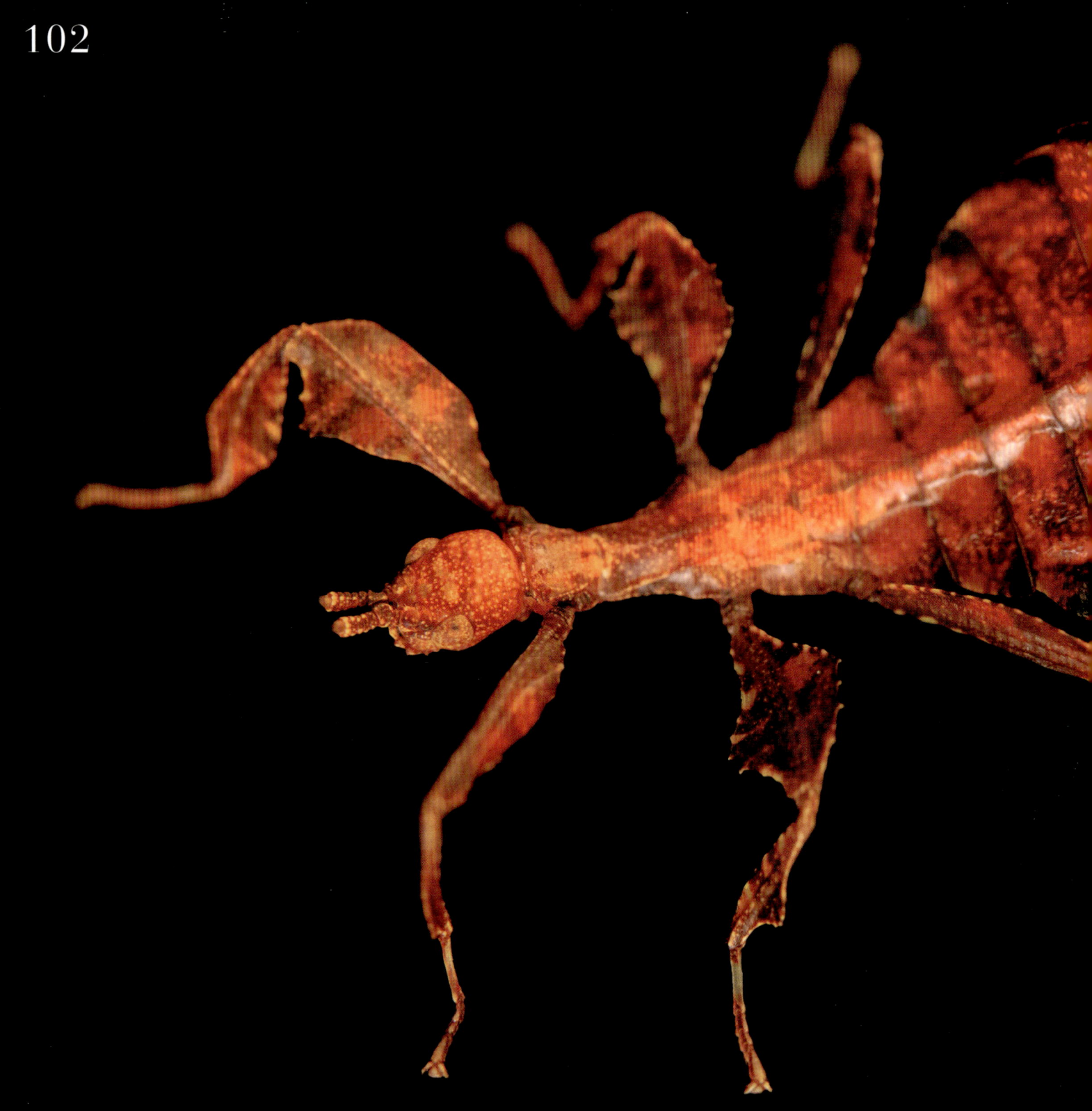

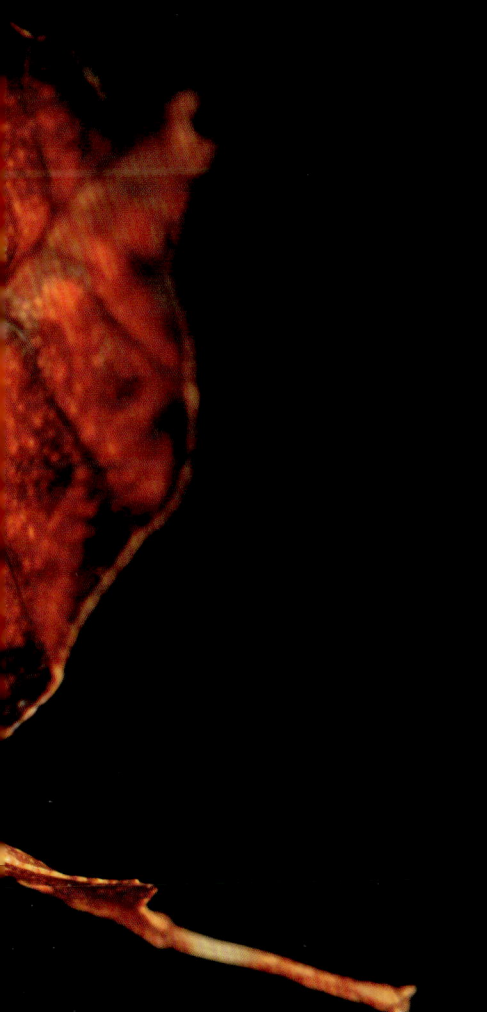

▼

Gray's leaf insect
Pulchriphyllium bioculatum (LC)

Although the youngest of these start out reddish or brown, they turn green once they start feeding on leaves, which helps them mimic their environment.

"For many species, nature gives babies every advantage it can, including the ability to hide—through camouflage or other ways of blending in."

Kagu
Rhynochetos jubatus (EN)

When kagu parents call their young, they sound like barking puppies. This chick will trade rust-colored down for pale gray adult plumage; its eyes will turn red, its legs a vivid orange.

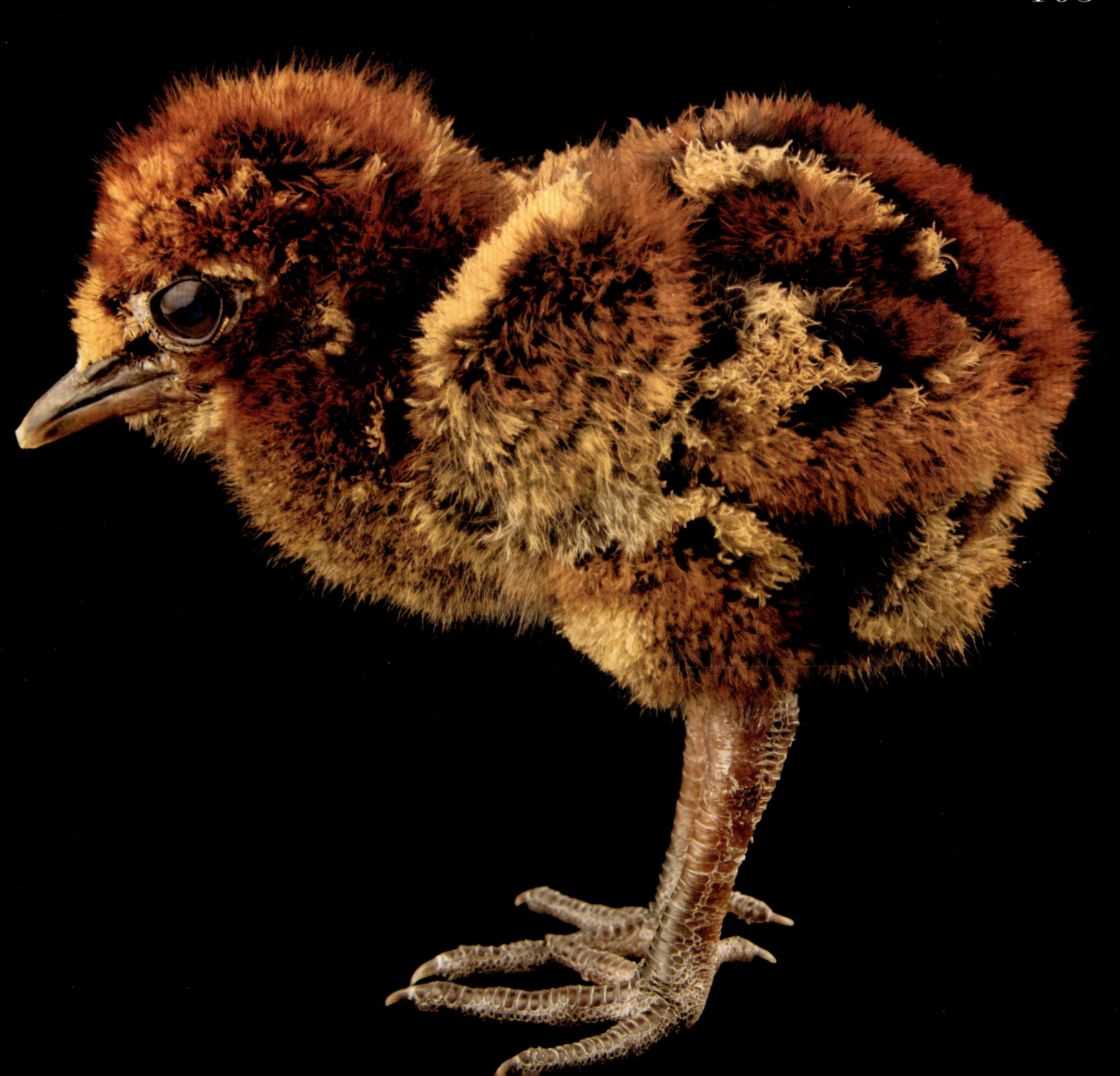

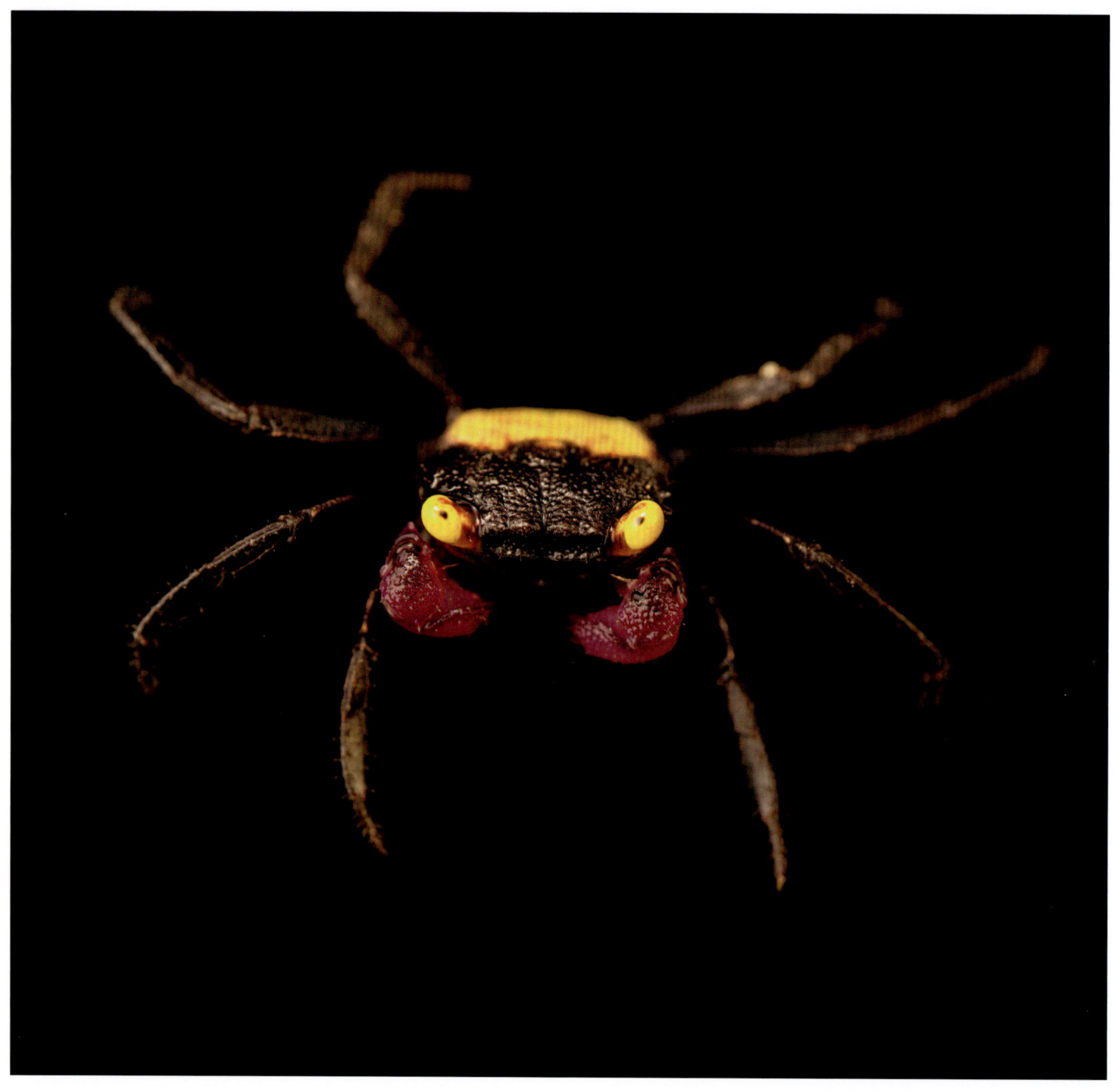

Purple vampire crabs

Geosesarma dennerle (NE)

These feisty nocturnal crabs of the forests, rivers, and lakes of Indonesia will be only two inches (5 cm) long when fully grown—and that includes their legs.

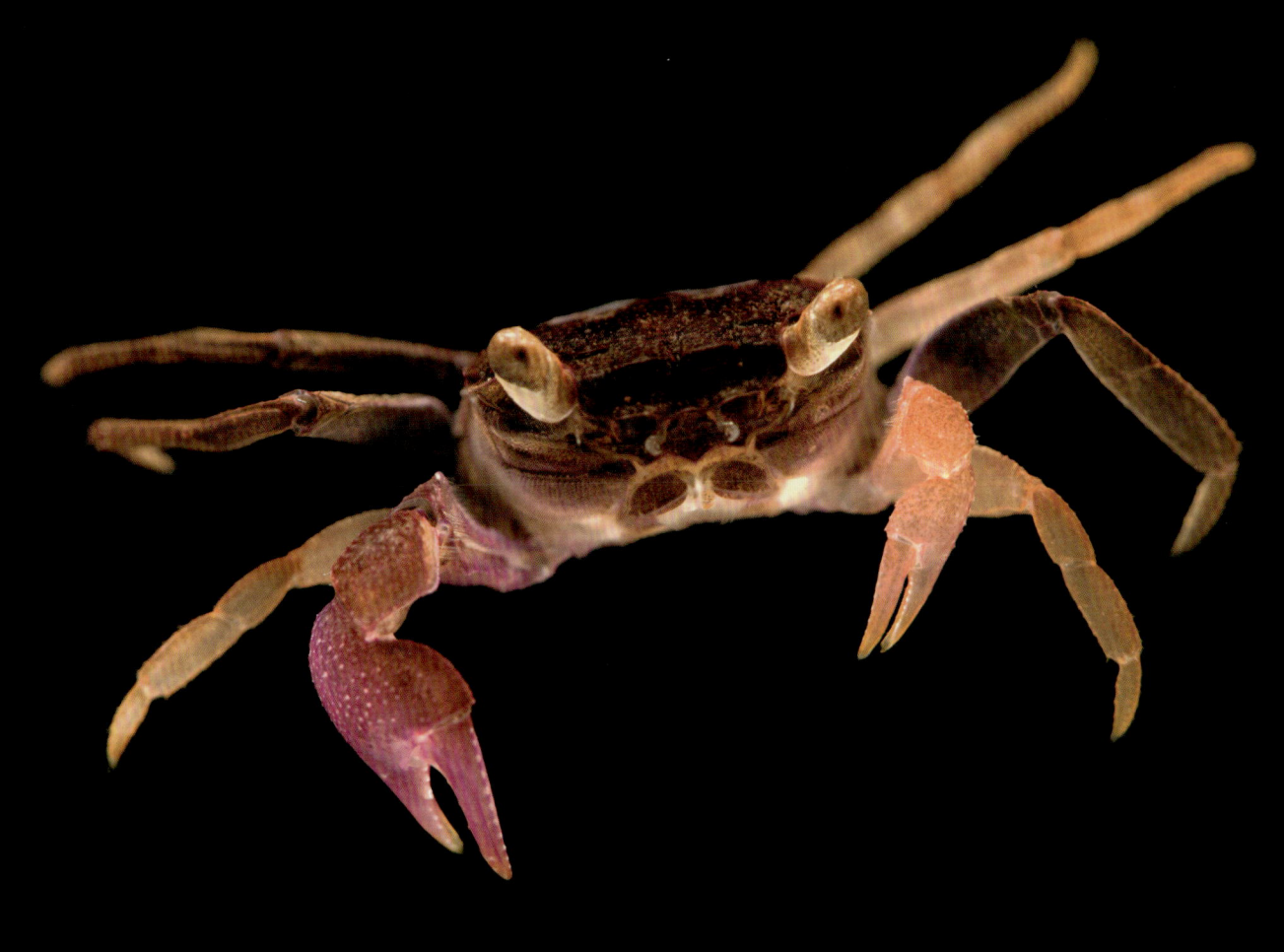

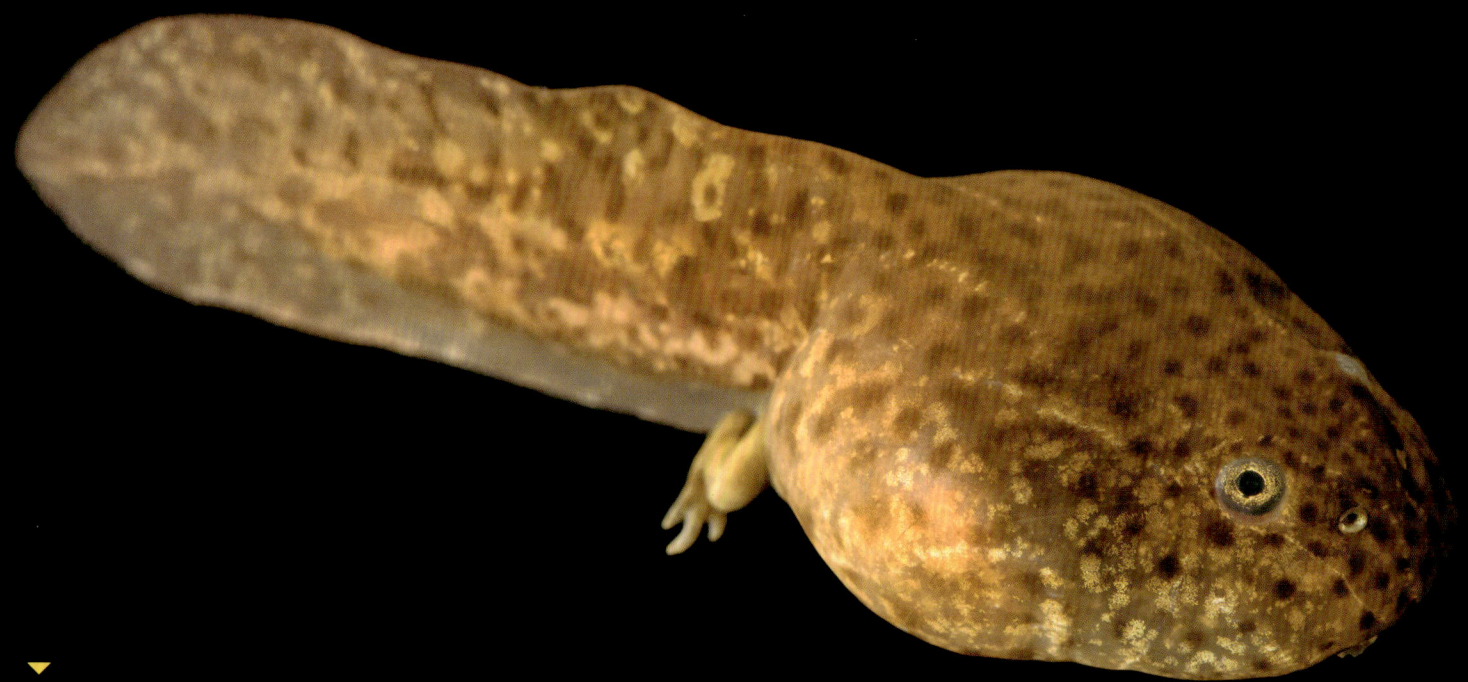

▼

Titicaca water frogs
Telmatobius culeus (EN)

Tadpoles, like the one above, hatch along
the shore of Lake Titicaca, a high-altitude
lake in Bolivia and Peru. A young frog (right)
remains in shallow water; adults, fully aquatic,
live deep underwater.

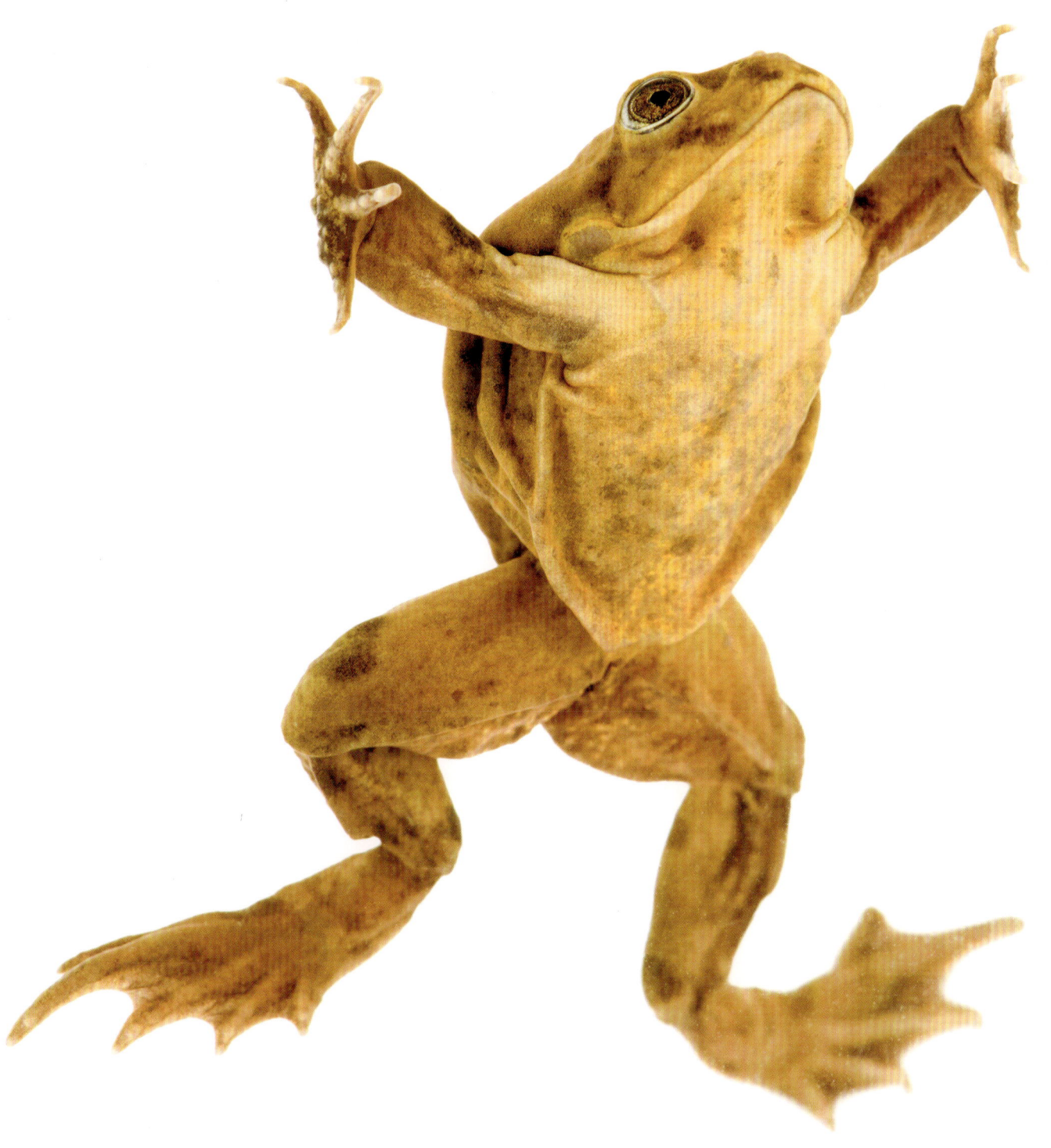

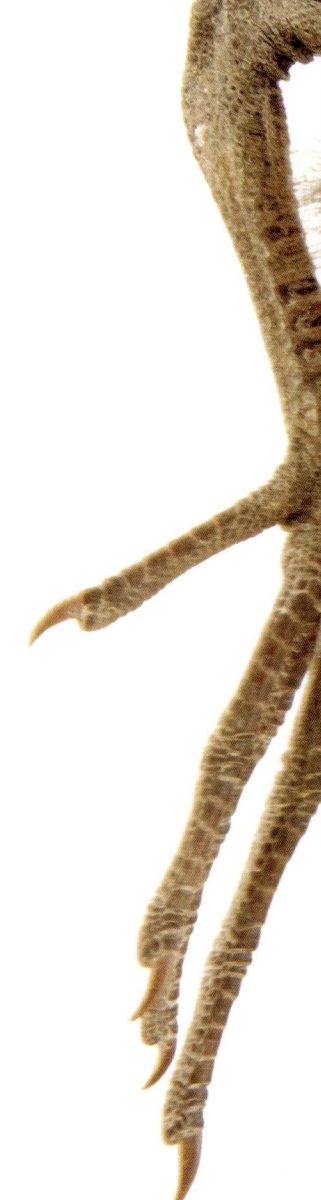

White-breasted waterhen

Amaurornis phoenicurus (LC)

White-breasted waterhen chicks start out as dark balls of fluff,
but by adulthood they'll sport their signature white face and front.
Long legs help them pick their way around marshes.

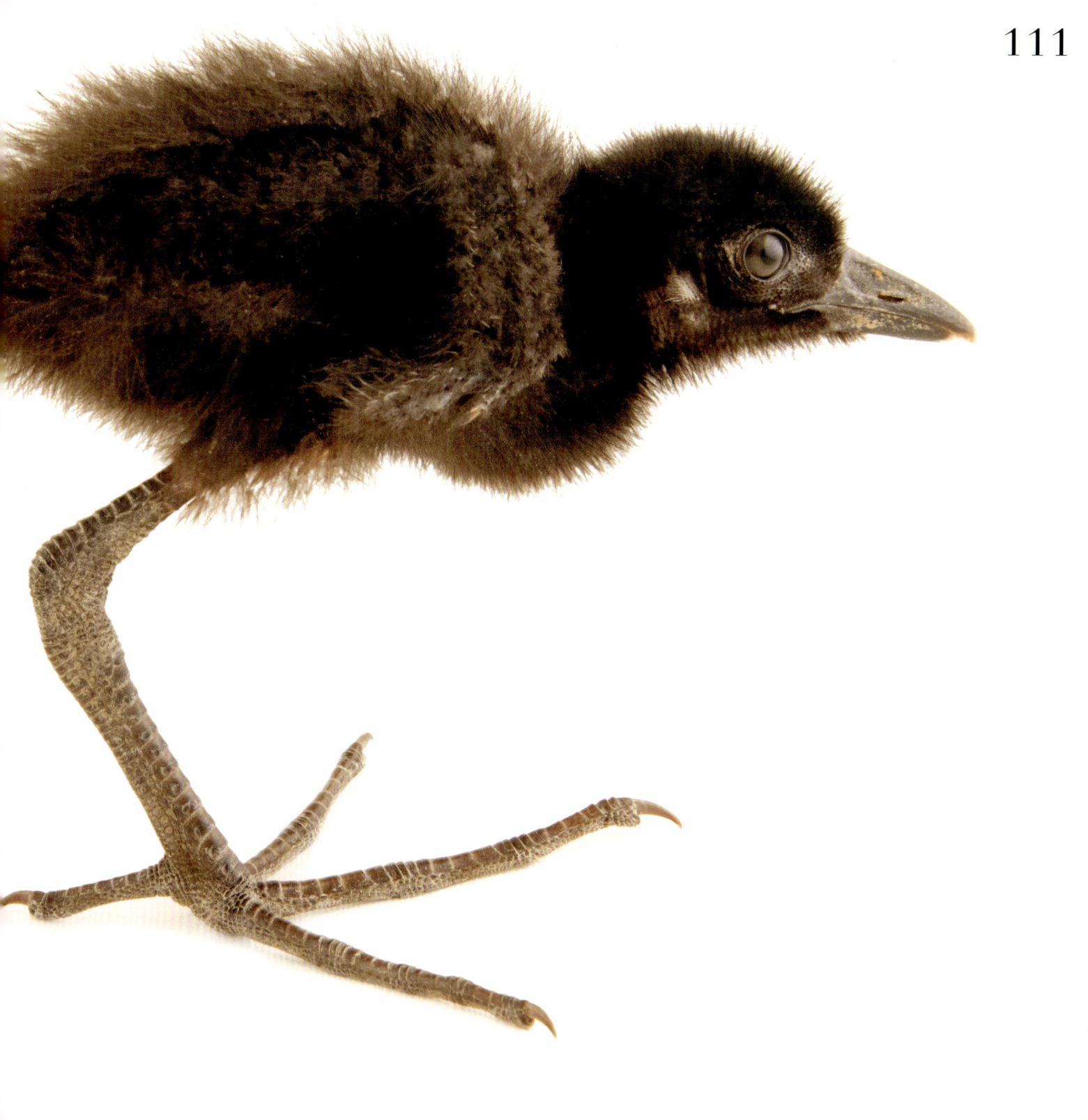

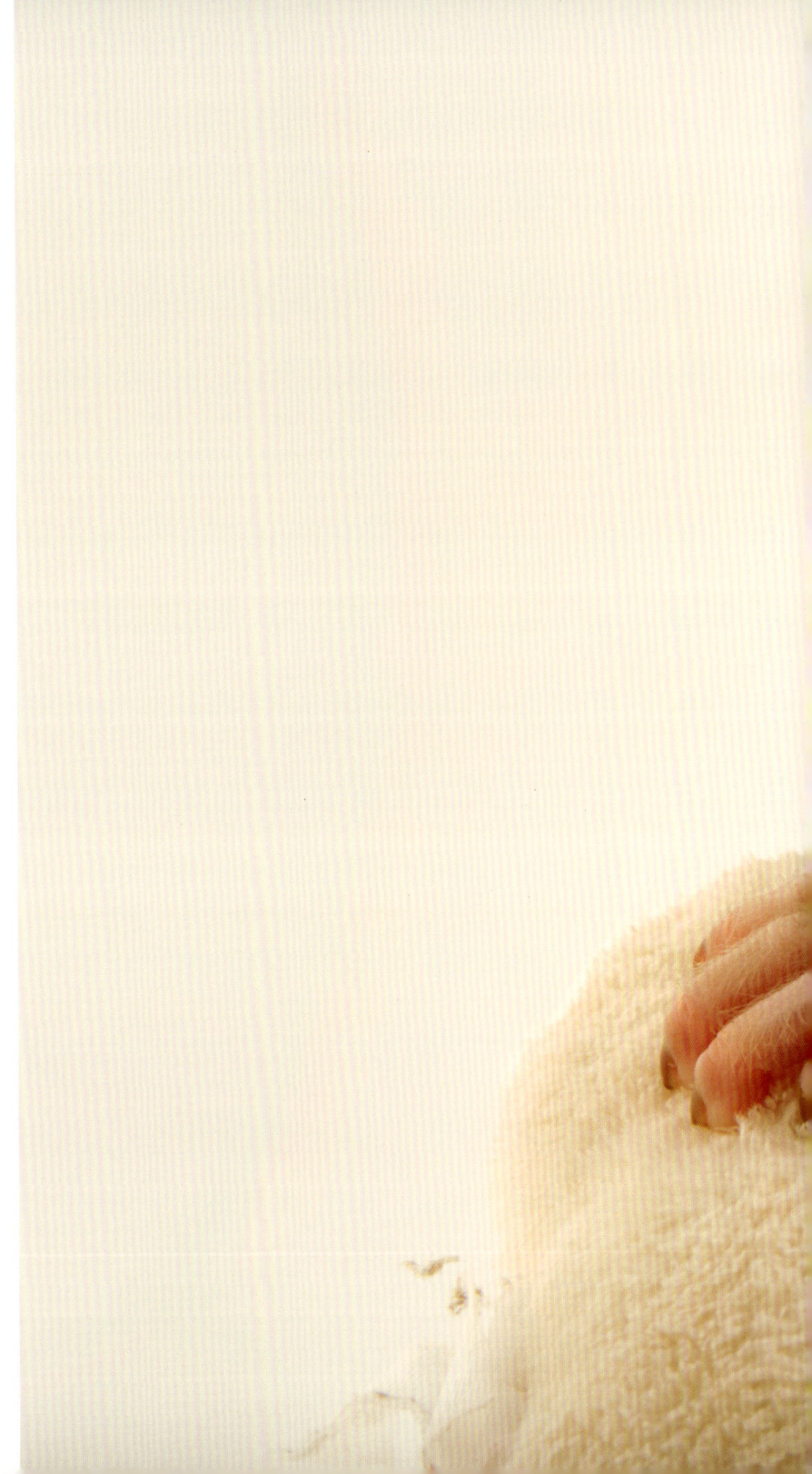

Japanese macaque

Macaca fuscata fuscata (LC)

Young Japanese macaques, or snow
monkeys, will frolic in freshly fallen
snow, rolling snowballs downhill
simply for the joy of it. This little one
is two weeks old.

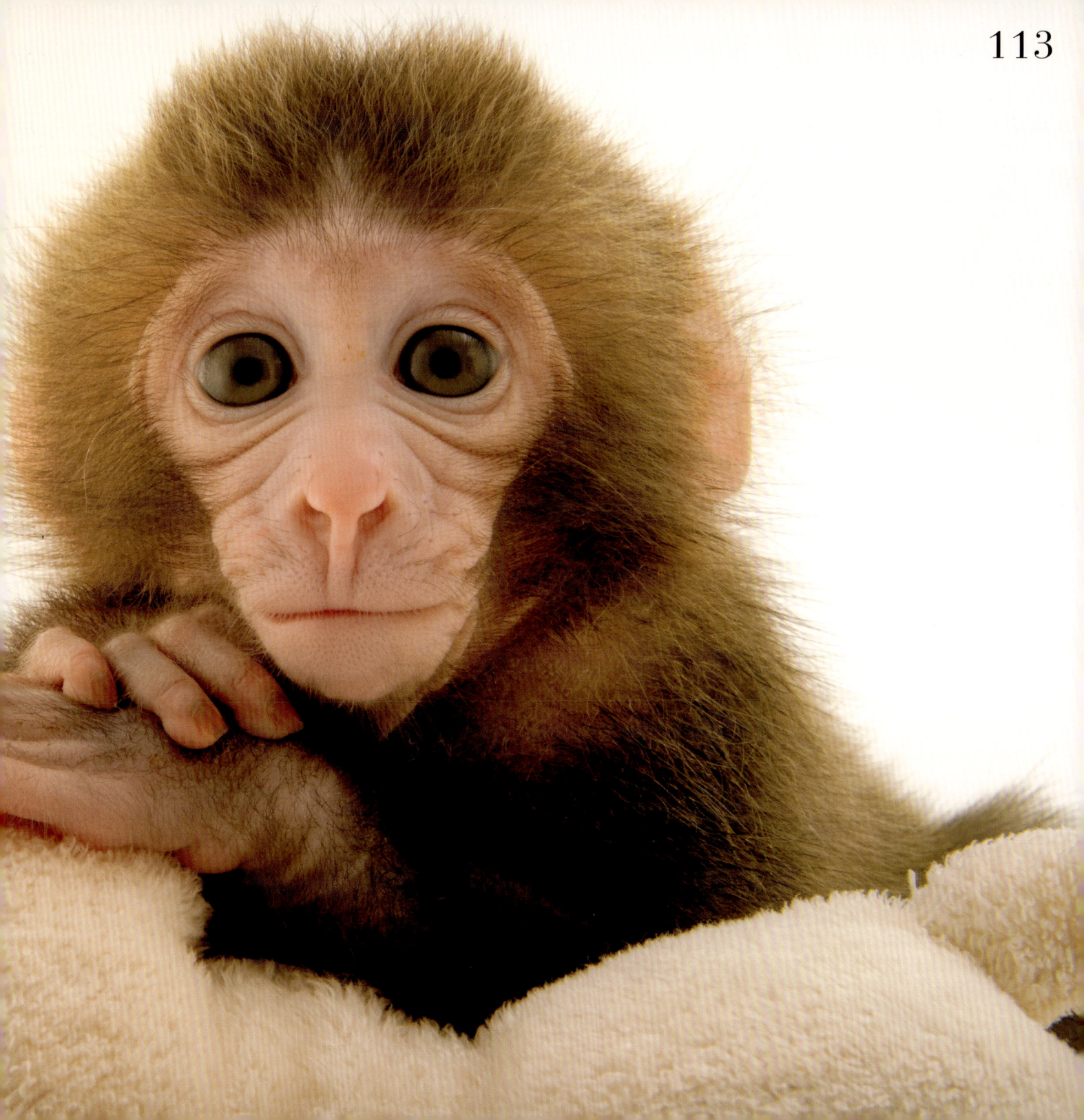

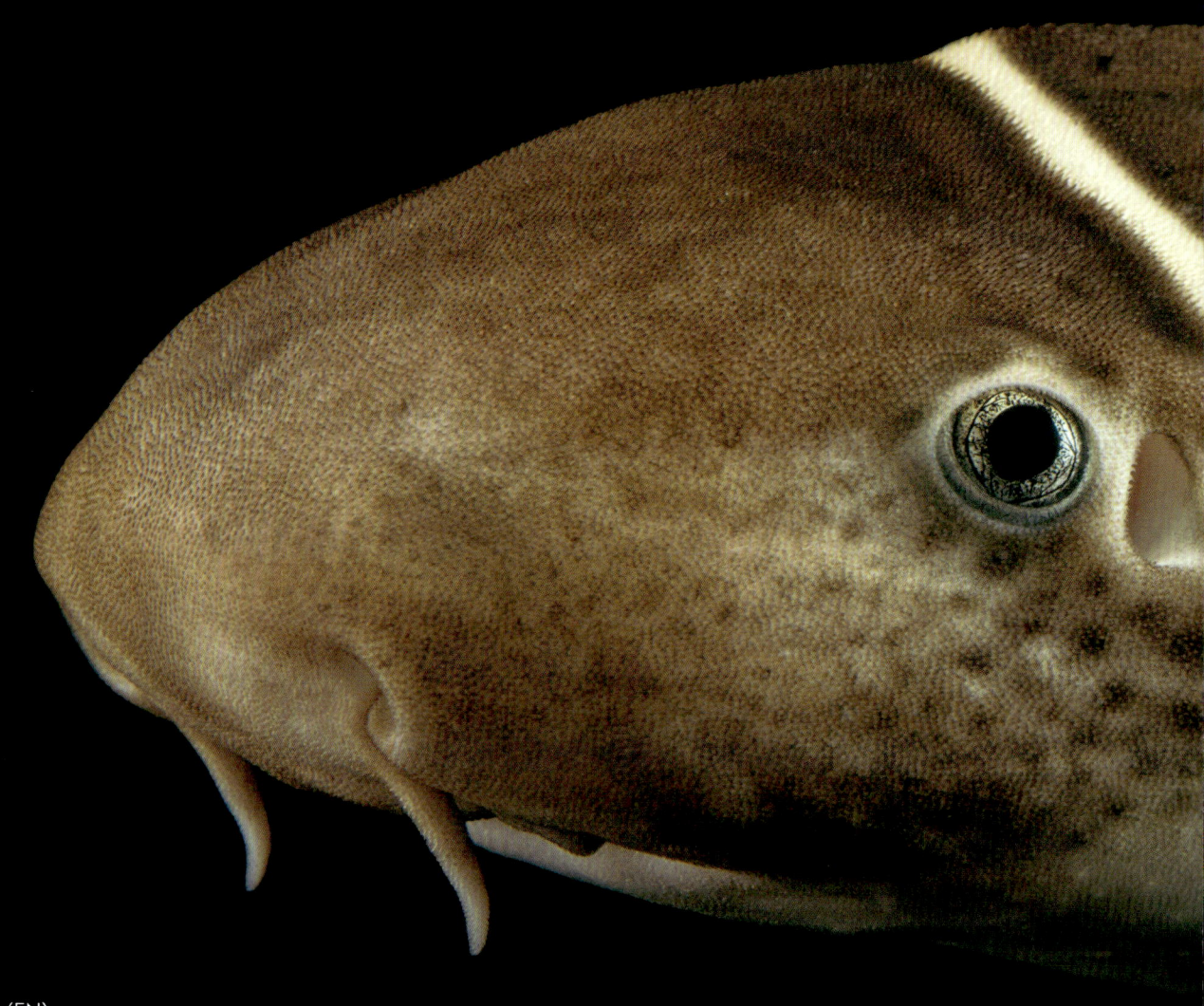

Zebra shark

Stegostoma tigrinum (EN)

This young shark's stripes are a defense
mechanism: They deter predators by
mimicking the appearance of a venomous
sea snake. Whisker-like barbels help
the nocturnal shark find prey in the dark.

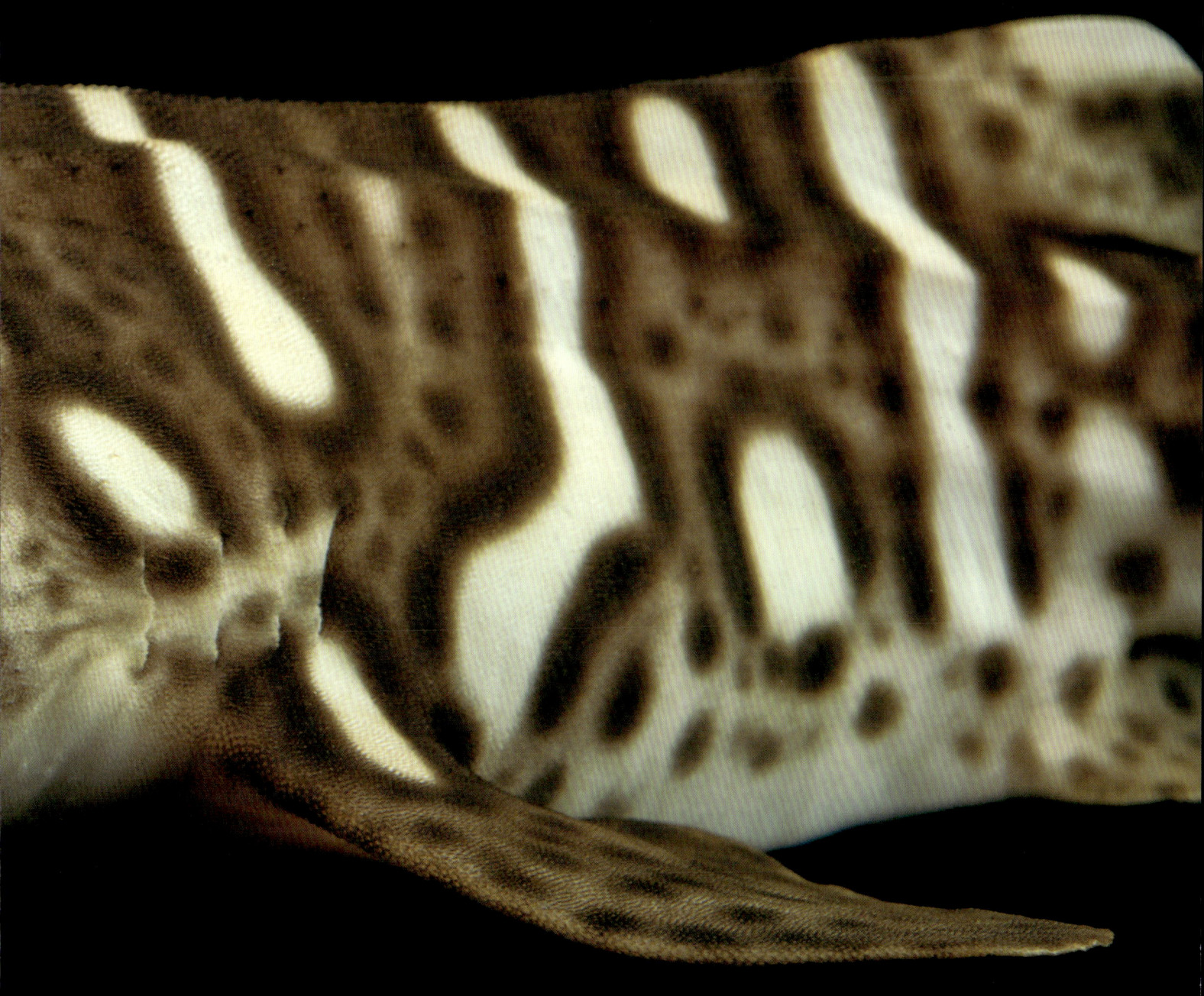

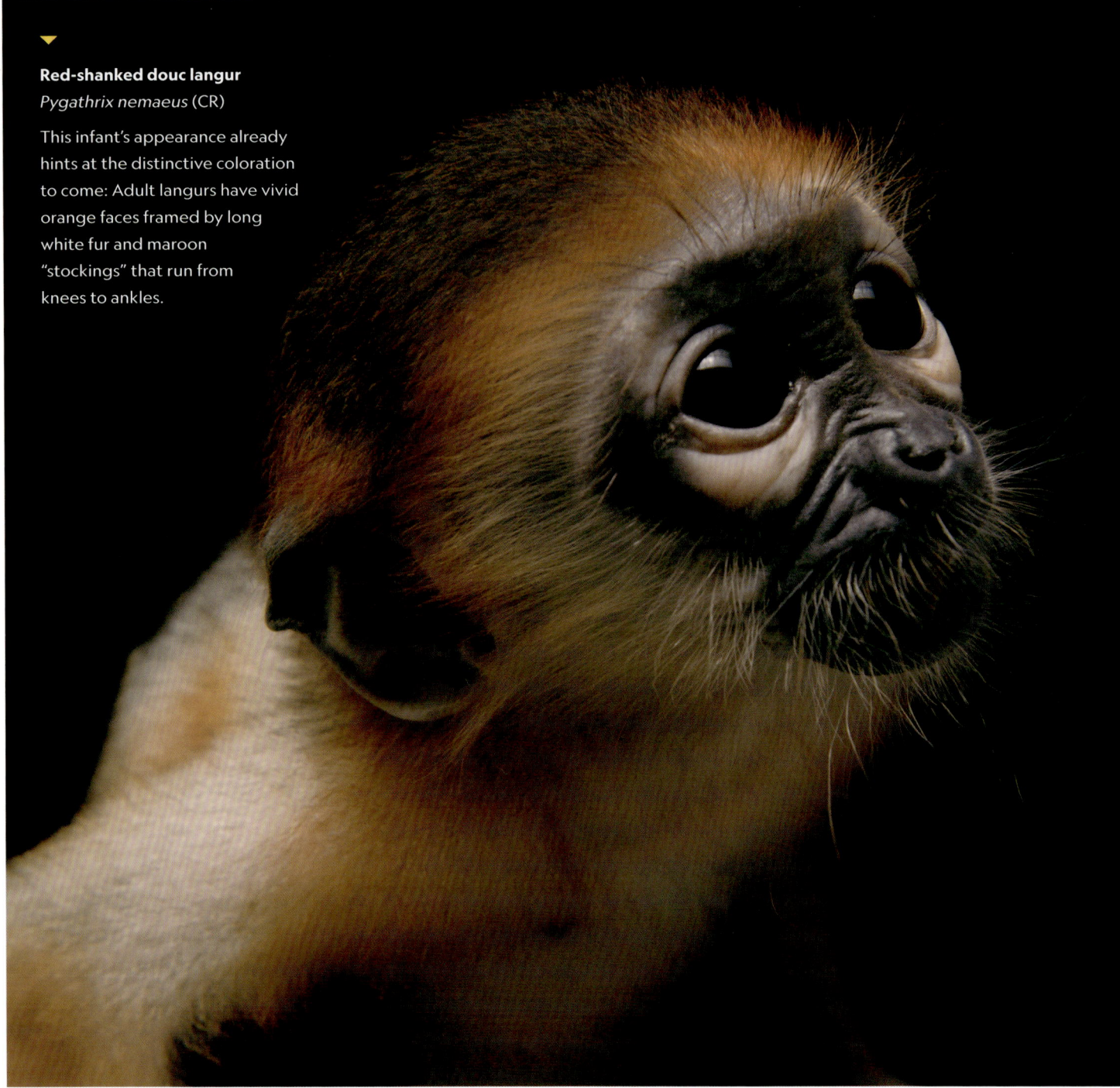

Red-shanked douc langur

Pygathrix nemaeus (CR)

This infant's appearance already hints at the distinctive coloration to come: Adult langurs have vivid orange faces framed by long white fur and maroon "stockings" that run from knees to ankles.

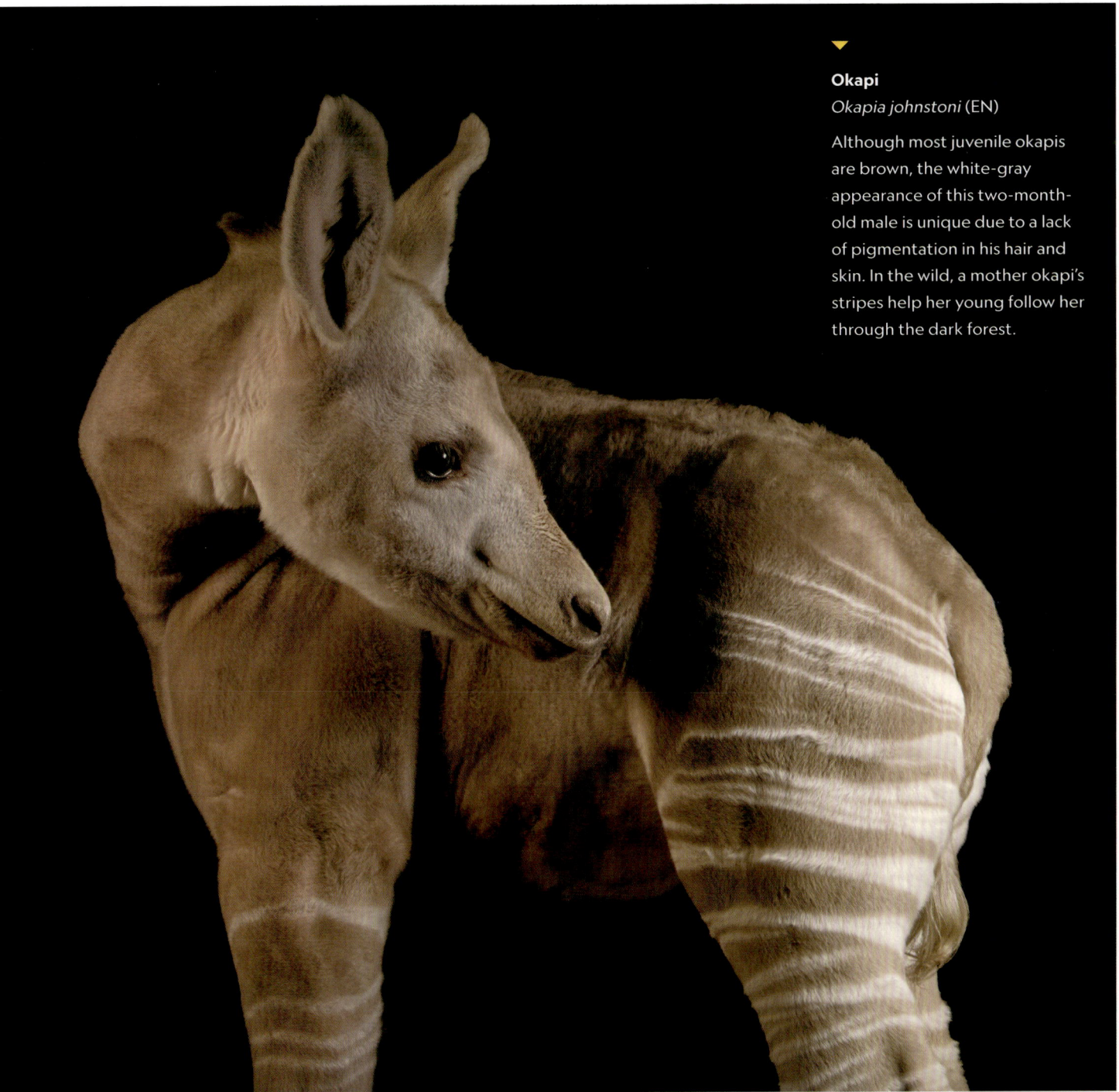

Okapi
Okapia johnstoni (EN)

Although most juvenile okapis are brown, the white-gray appearance of this two-month-old male is unique due to a lack of pigmentation in his hair and skin. In the wild, a mother okapi's stripes help her young follow her through the dark forest.

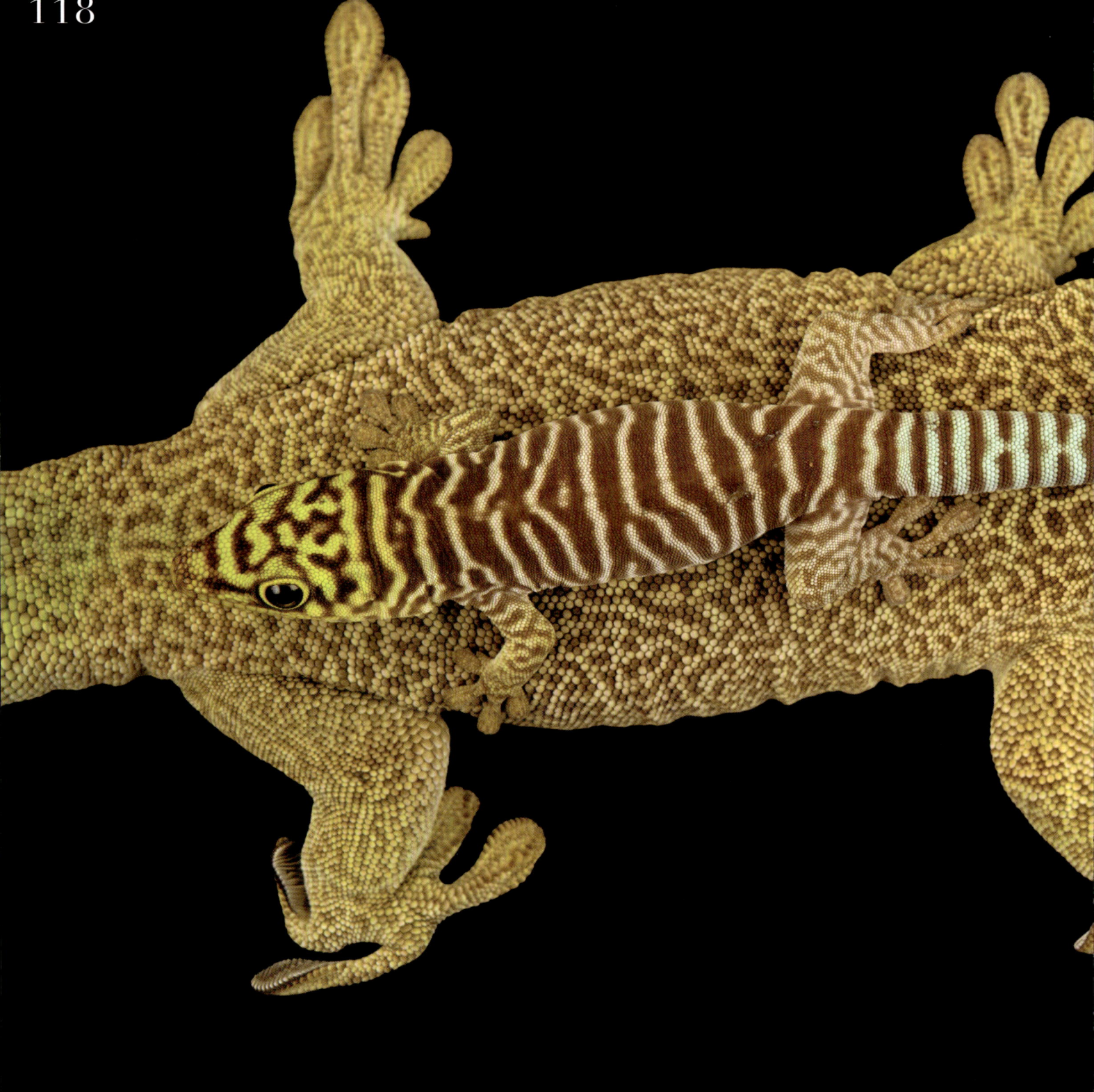

Standing's day geckos
Phelsuma standingi (VU)

This colorful, tree-dwelling lizard native to southwestern Madagascar is one of the largest living species of day gecko. Its population is declining in part because it is harvested for the international pet trade.

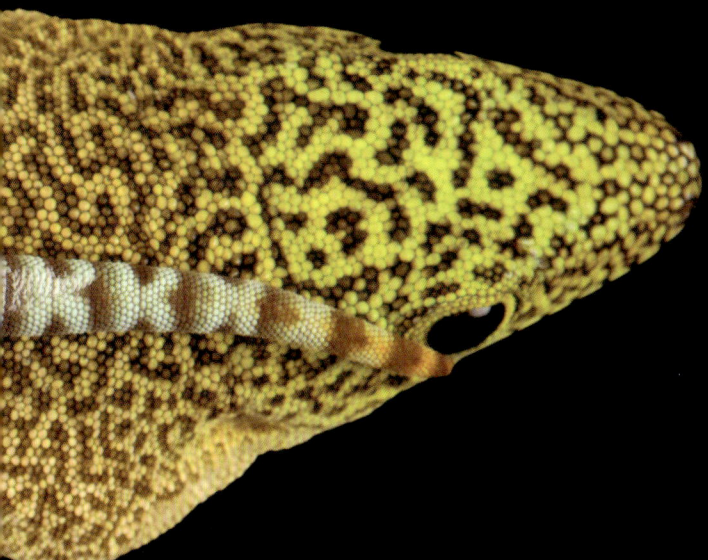

"The Photo Ark is my last-ditch effort to try to get the public to care about species and habitats while there is still time to save them."

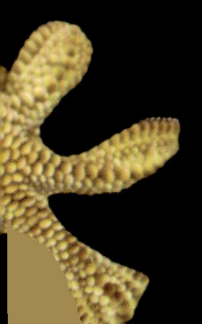

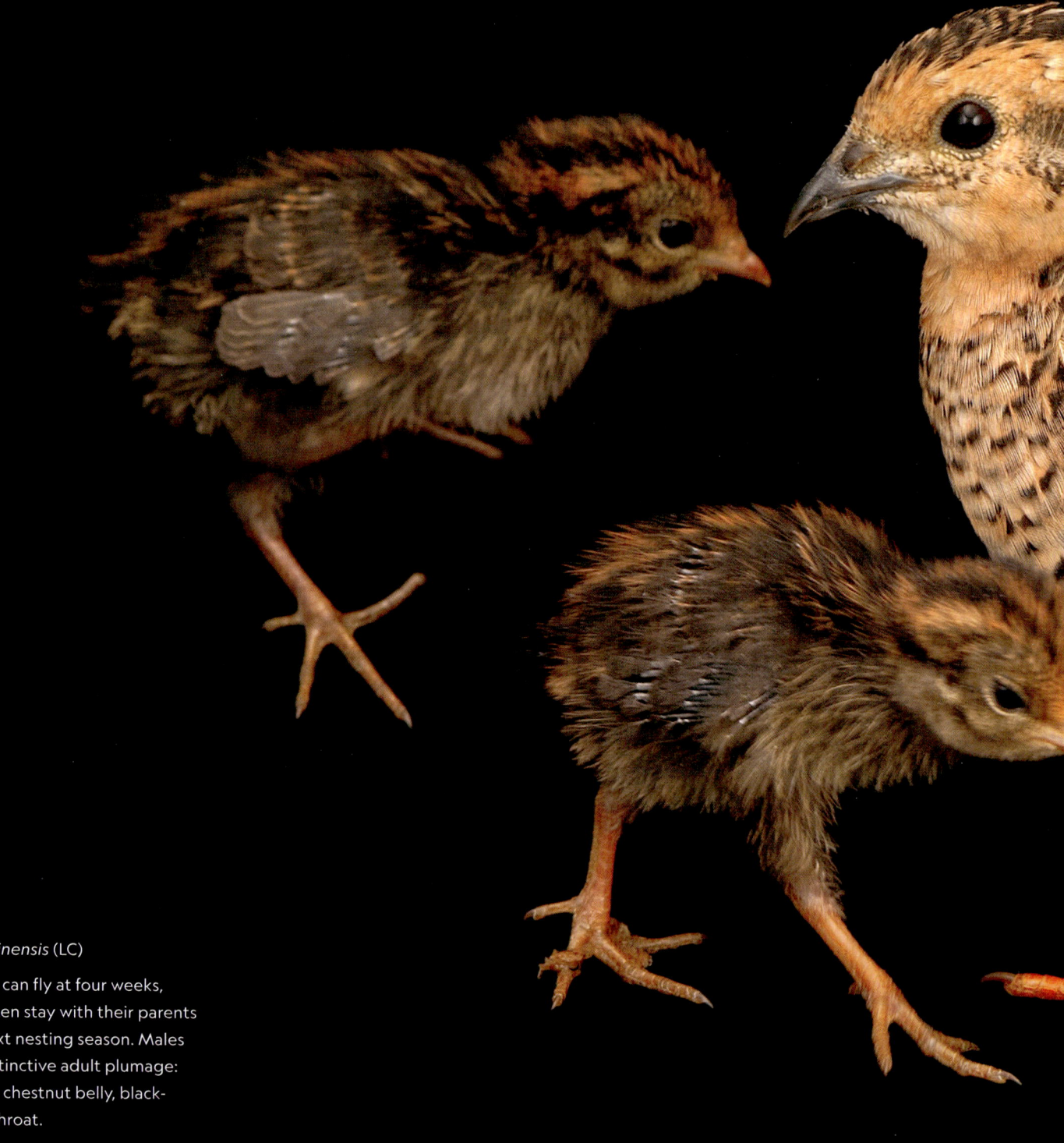

King quails

Synoicus chinensis (LC)

Quail chicks can fly at four weeks,
but they often stay with their parents
until the next nesting season. Males
develop distinctive adult plumage:
blue breast, chestnut belly, black-
and-white throat.

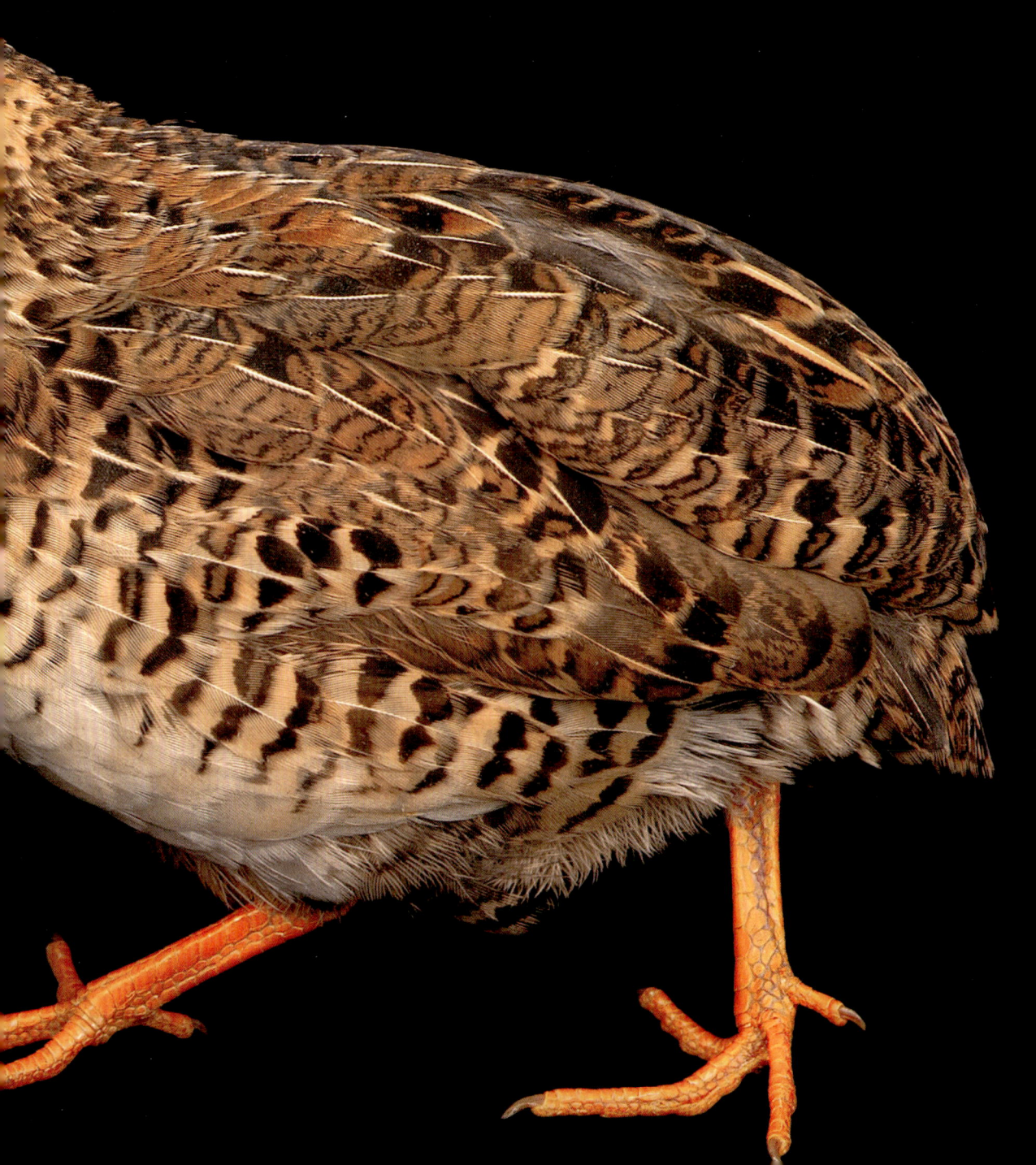

Snake River sockeye salmon
Oncorhynchus nerka (LC)

This subspecies of salmon native to western North America is believed to travel the farthest inland and to spawn at the highest altitude of any sockeye salmon. These fry, just a couple months old, began their lives far upstream.

BEHIND THE LENS

CIRCLE OF LIFE

ANIMAL: Snake River sockeye salmon, *Oncorhynchus nerka* (LC)

LOCATION: Eagle Fish Hatchery, Eagle, Idaho

"Salmon have an amazing story. They are born in a stream. The babies stay upriver for a time, getting bigger. Then they let the current wash them out to sea. They become oceangoing creatures and live in the sea until it's time to spawn. At that time, they find their way back upstream and lay their eggs. Then they die. Why? Because they're feeding their young. Their decaying bodies allow for a bloom of aquatic invertebrates, which feed on rotting flesh and plants growing thanks to nutrients flooding the water. It all provides food for the hatching salmon eggs. If it weren't for the bodies of their parents creating this abundance of food, the young salmon fry would have nothing to eat. The parents' bodies create a grocery store for the babies."

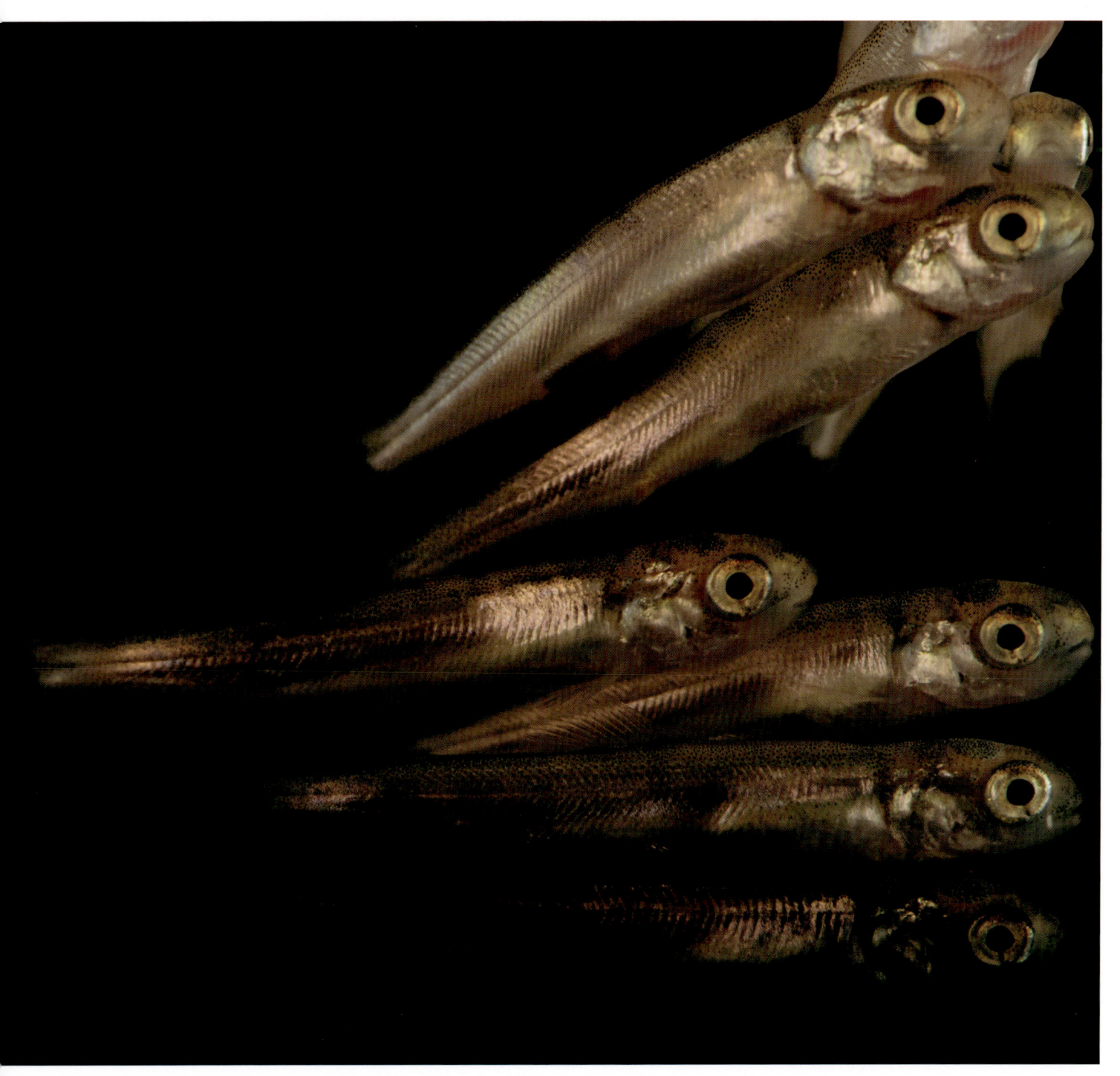

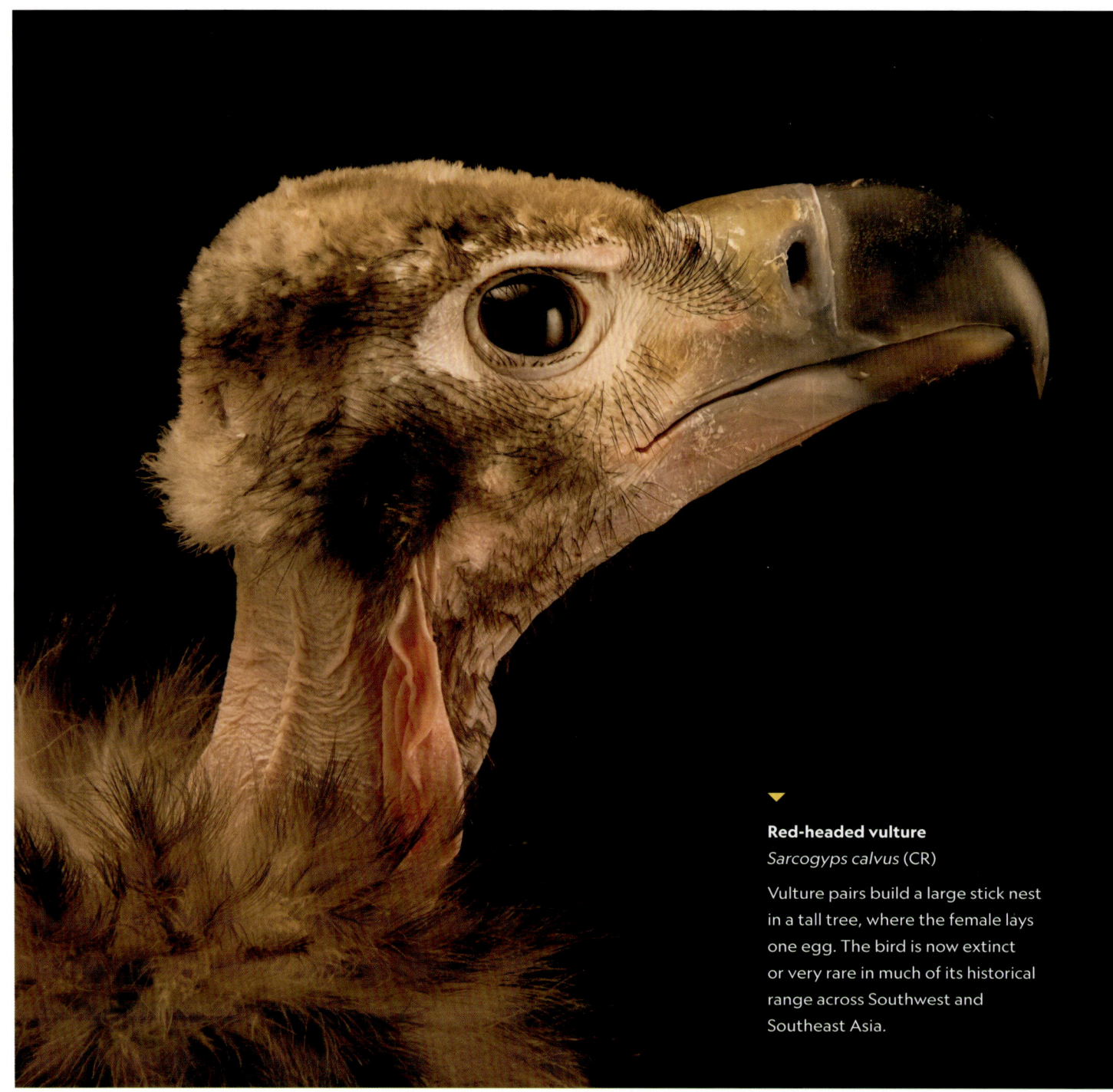

Red-headed vulture
Sarcogyps calvus (CR)

Vulture pairs build a large stick nest
in a tall tree, where the female lays
one egg. The bird is now extinct
or very rare in much of its historical
range across Southwest and
Southeast Asia.

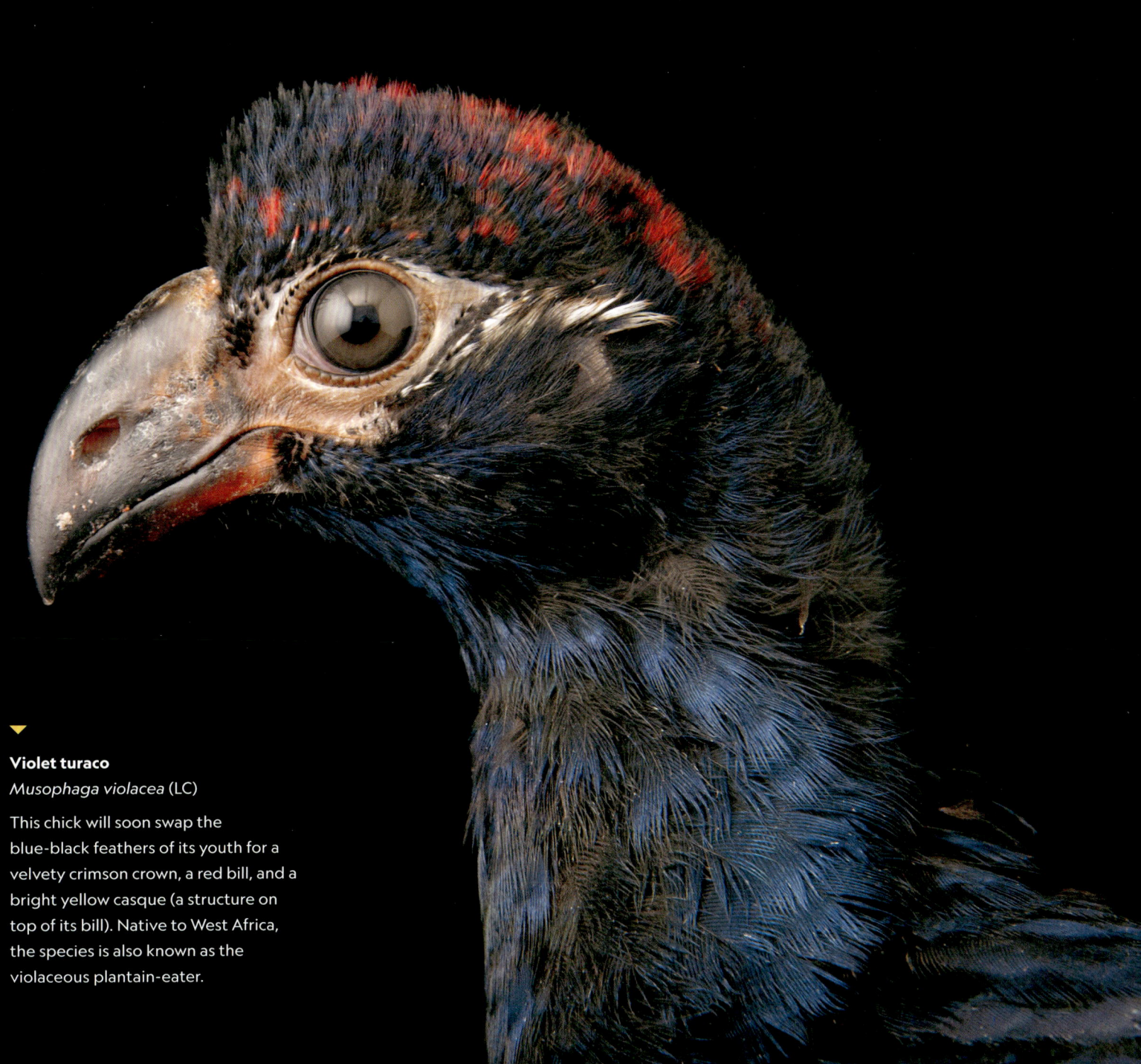

Violet turaco

Musophaga violacea (LC)

This chick will soon swap the
blue-black feathers of its youth for a
velvety crimson crown, a red bill, and a
bright yellow casque (a structure on
top of its bill). Native to West Africa,
the species is also known as the
violaceous plantain-eater.

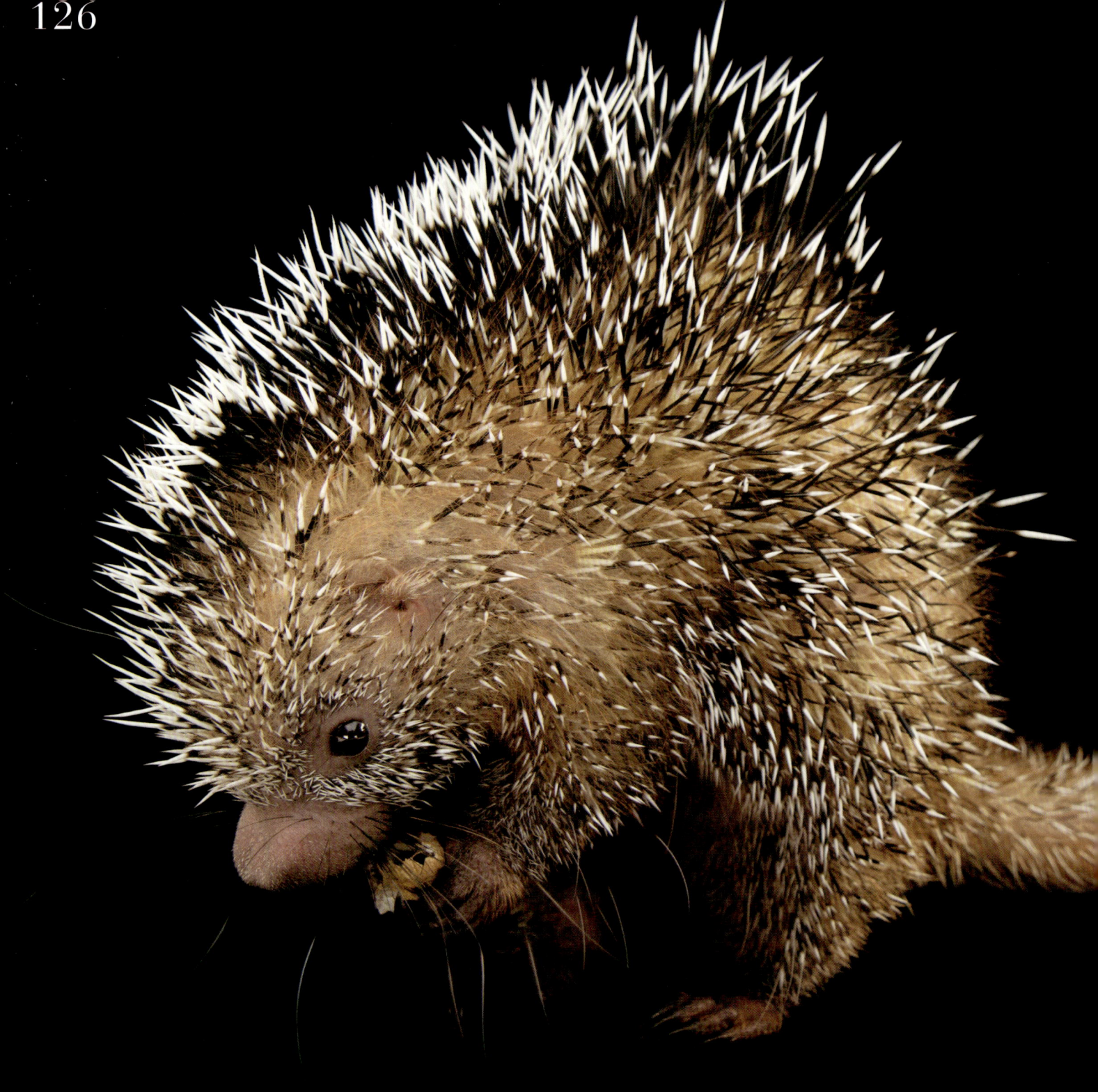

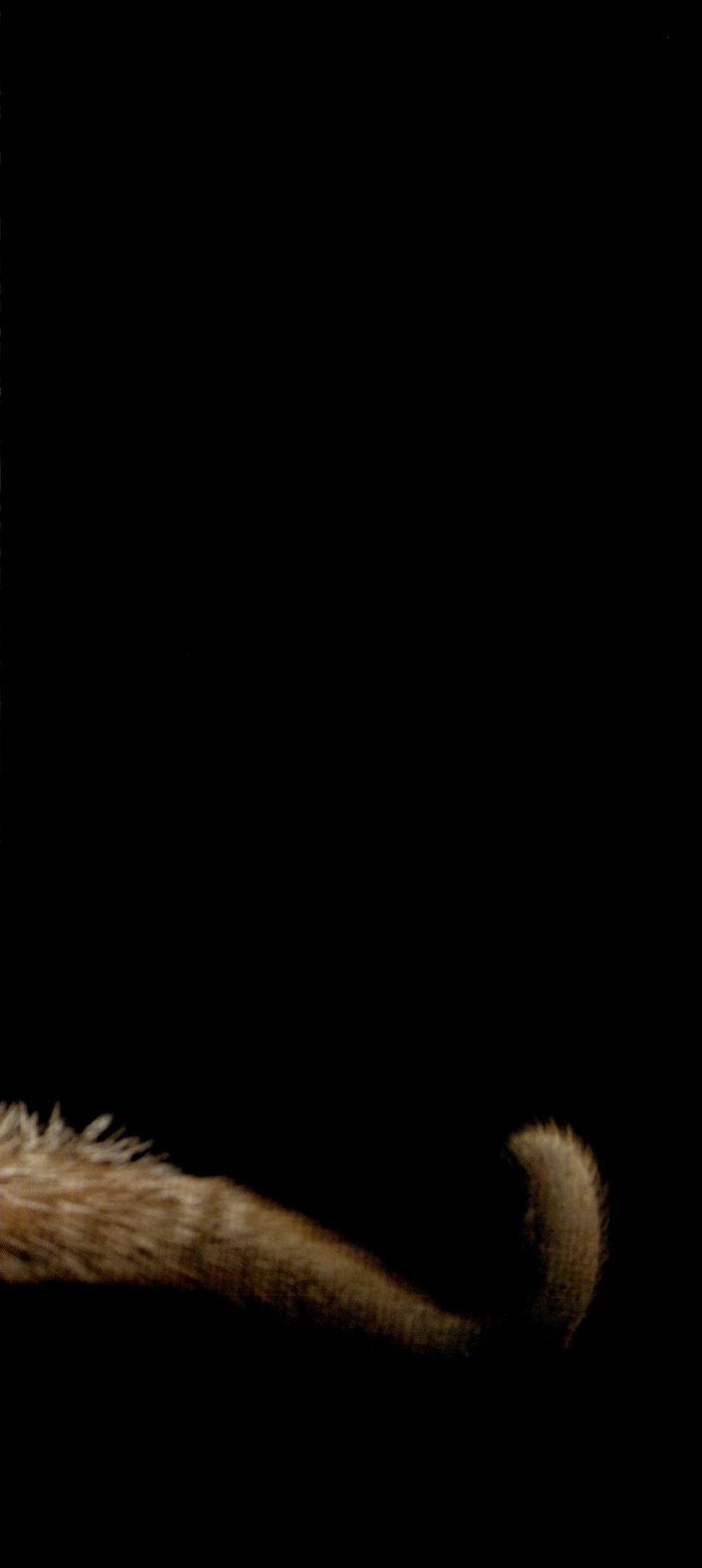

Brazilian porcupine

Coendou prehensilis (LC)

These porcupine babies are born with a thick coat of hairs that harden into sharp quills. A long, strong tail helps them move from tree to tree in the lowland forests of South America.

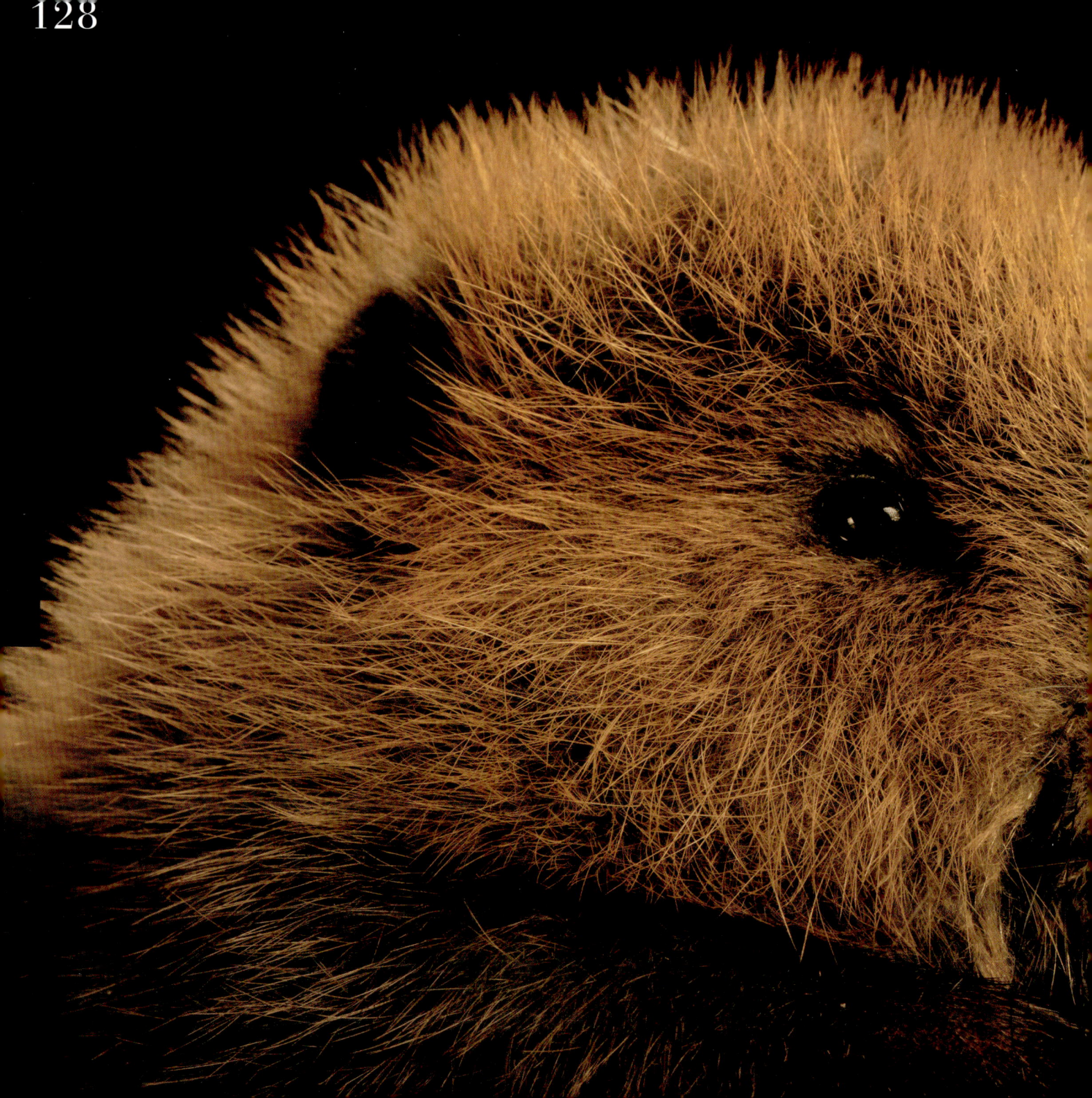

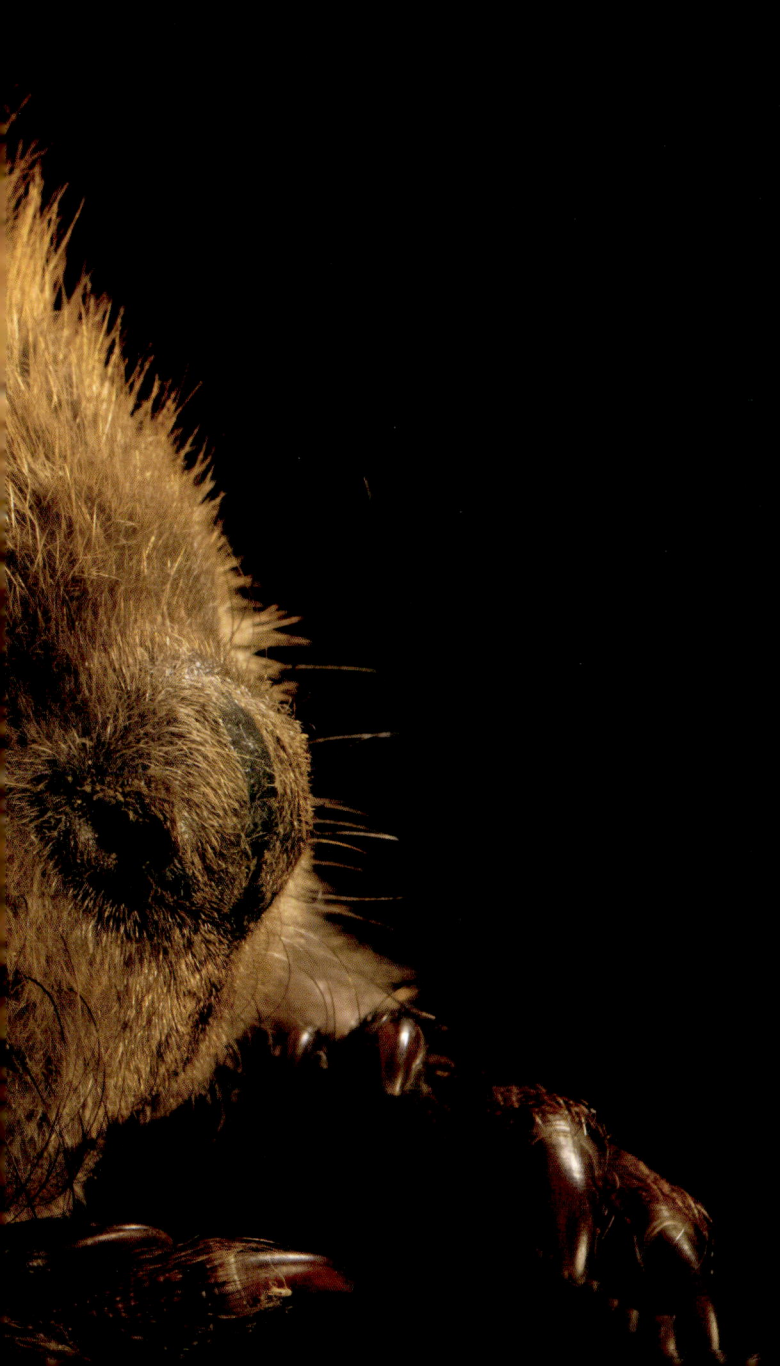

▼

North American beaver
Castor canadensis (LC)

Beaver kits like this two-week-old youngster learn early how to be savvy engineers. By felling trees and building dams, canals, and lodges, they have a huge impact on their wetland homes.

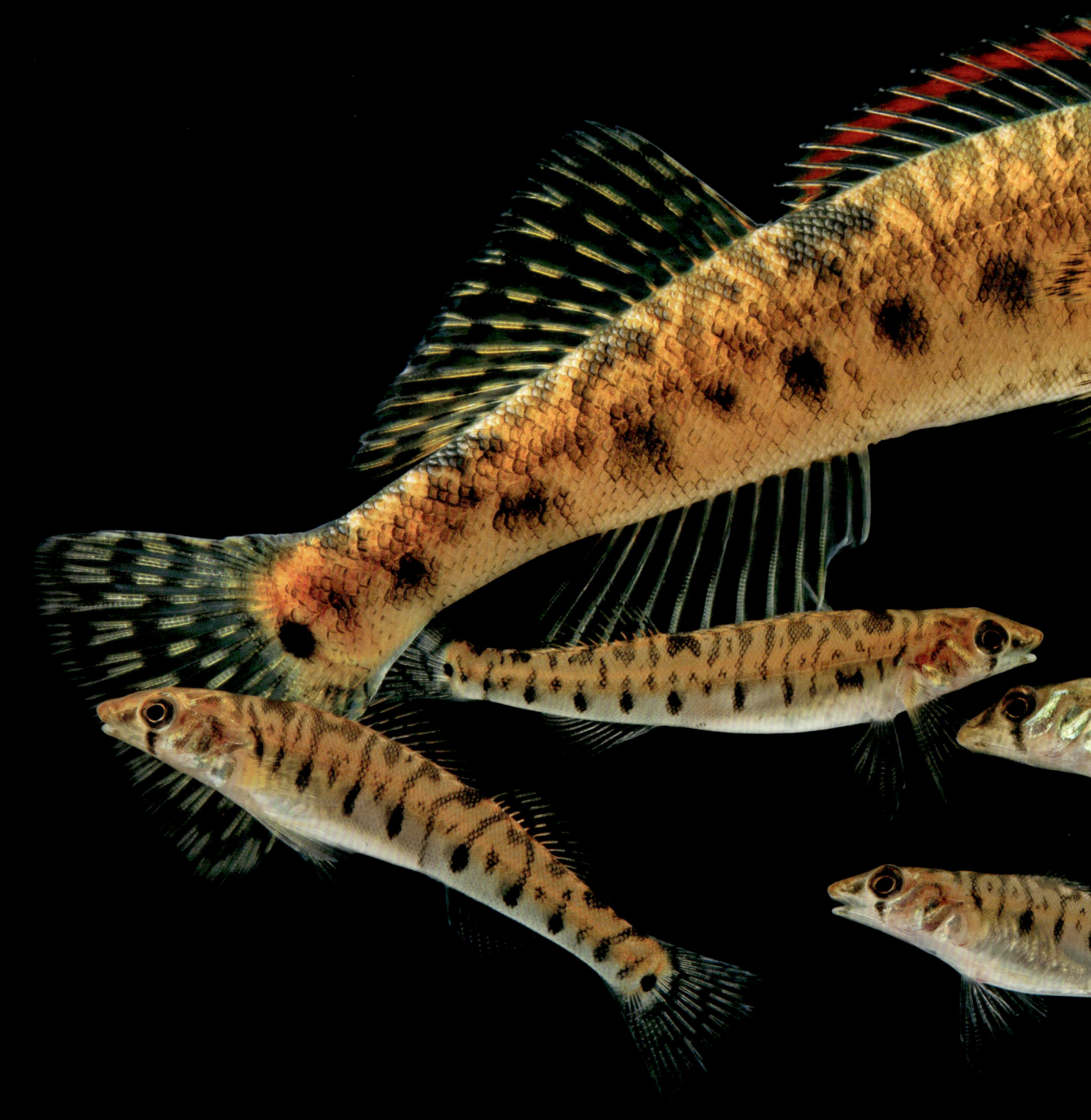

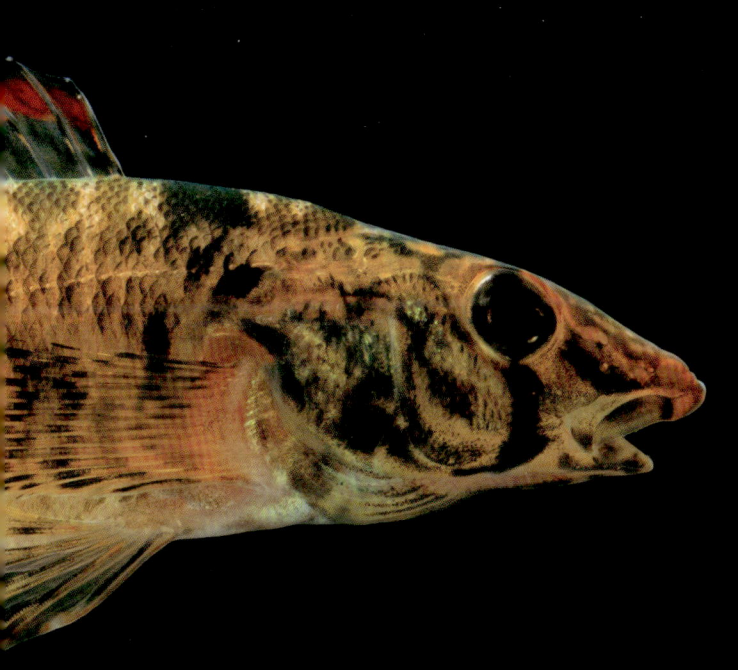

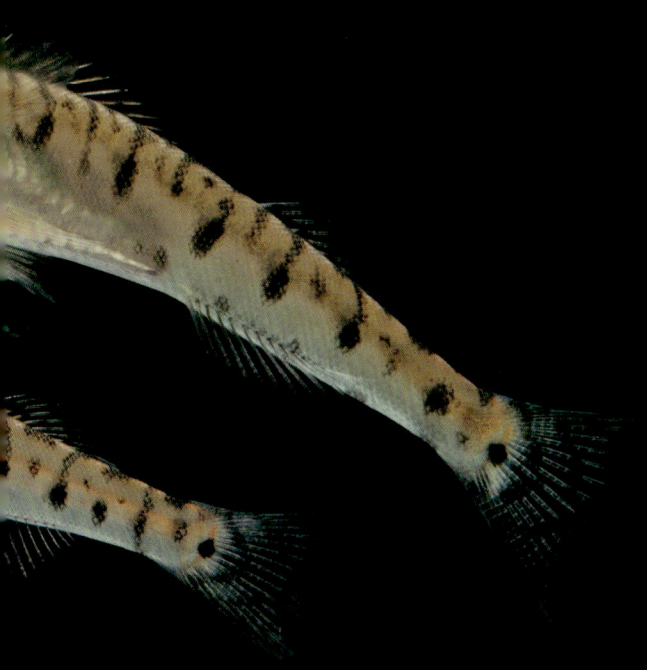

4 | FAMILY

ALL IN THIS TOGETHER

What is family to a seabird who soars alone over the open ocean,
or to a primate who clings to its mother for years? Do animals have siblings,
aunties, or cousins? Can they survive without any family at all?

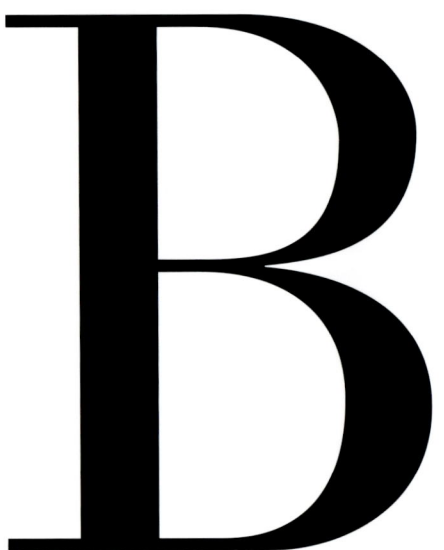

BEARING AND REARING young are central to the life cycle of nearly all creatures, whether done the way of the giant clam—by releasing half a billion eggs into the vast sea, full stop—or the koala, who provides long stretches of intense care. Parents of the animal kingdom are endlessly imaginative in how they feed, shelter, and teach their youngsters. There is no one way, they might counsel us, to raise a tadpole or an owlet.

Females often assume the primary parenting duties, but there are exceptions. The male Darwin's rhea, a large, flightless bird of South America, for instance, is a do-it-all dad. He builds a nest and incubates up to 30 eggs laid by different females, then raises the chicks to adulthood, teaching them survival skills and protecting them from predators. At the other end of the caregiving spectrum are young Komodo dragons. They spend their first year of life in trees to avoid their massive landbound parents, who have a nasty inclination to eat their young.

Siblings, other relatives, and adults in the community are sometimes on the scene for companionship or care. They fill a need for babysitting or socialization and, like the pelagic cormorants who help arrange nest material for their future siblings, provide a hand (or beak). A collaborative behavior known as alloparenting allows some adult animals to leave their little ones in the care of other grown-ups while they set out to find food: Some penguin species form nursery systems for their chicks because both parents often go to sea together for a day or more to catch fish. But other families, like that of the solitary clouded leopard, make it work without any help at all.

Family can be loving and sweet. Our hearts flip at a flotilla of ducklings paddling furiously behind their mom, or orphaned orangutans in a tight embrace, holding fast to the only family they know. Family ties are the key to survival—not just for the babies, but also for the species. With so much at stake, many adult animals muster their fiercest behavior to defend their young. When a wolf or bear draws near a muskox herd, the adult oxen turn head-on to show their sharp horns to the intruders. There on the Arctic tundra, they form a defensive circle around their babies, who huddle together inside the furry fortress, safe for another day.

▼

Indian rhinoceroses *Rhinoceros unicornis* (VU)
Previous pages: **Roanoke logperch** *Percina rex* (VU)

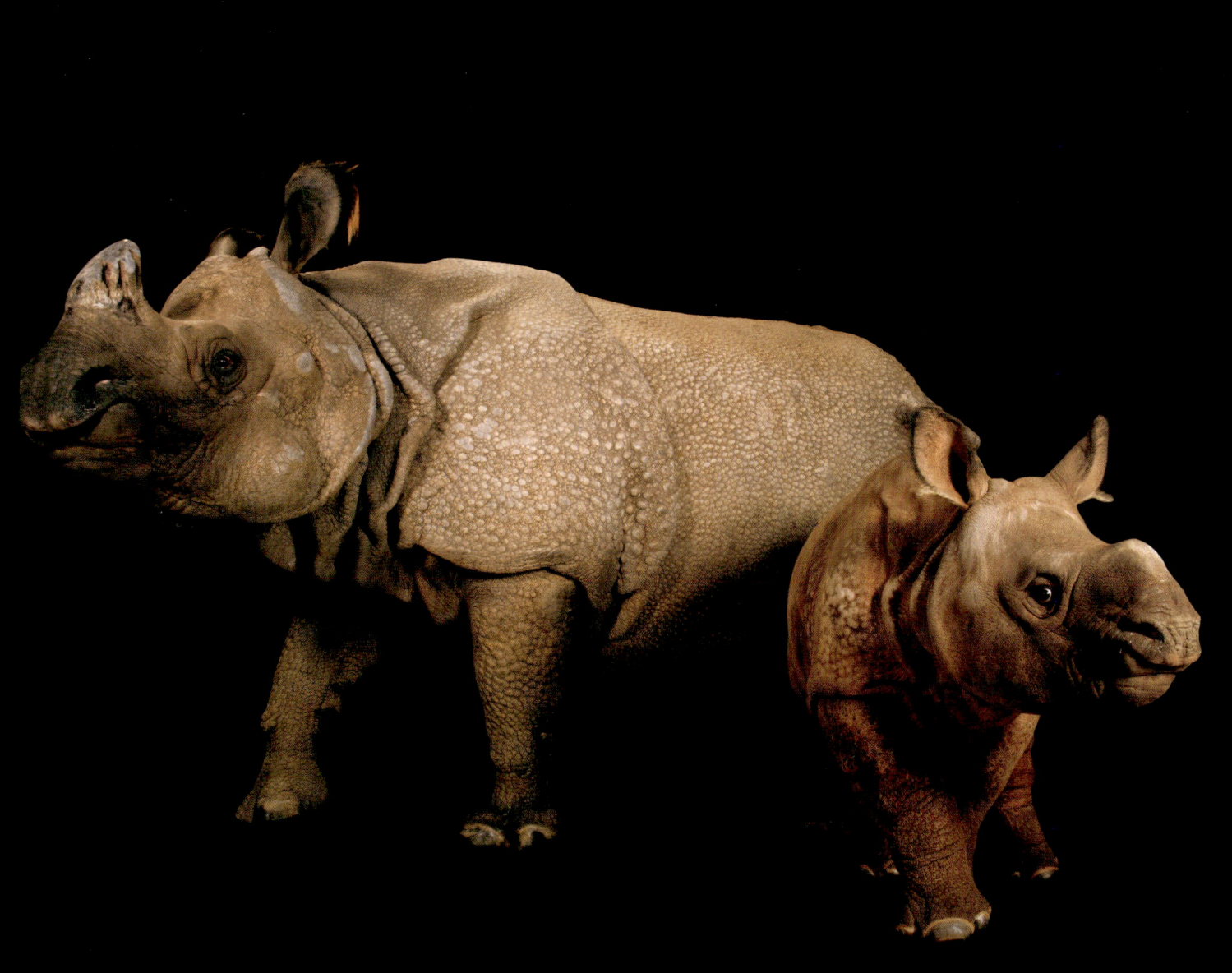

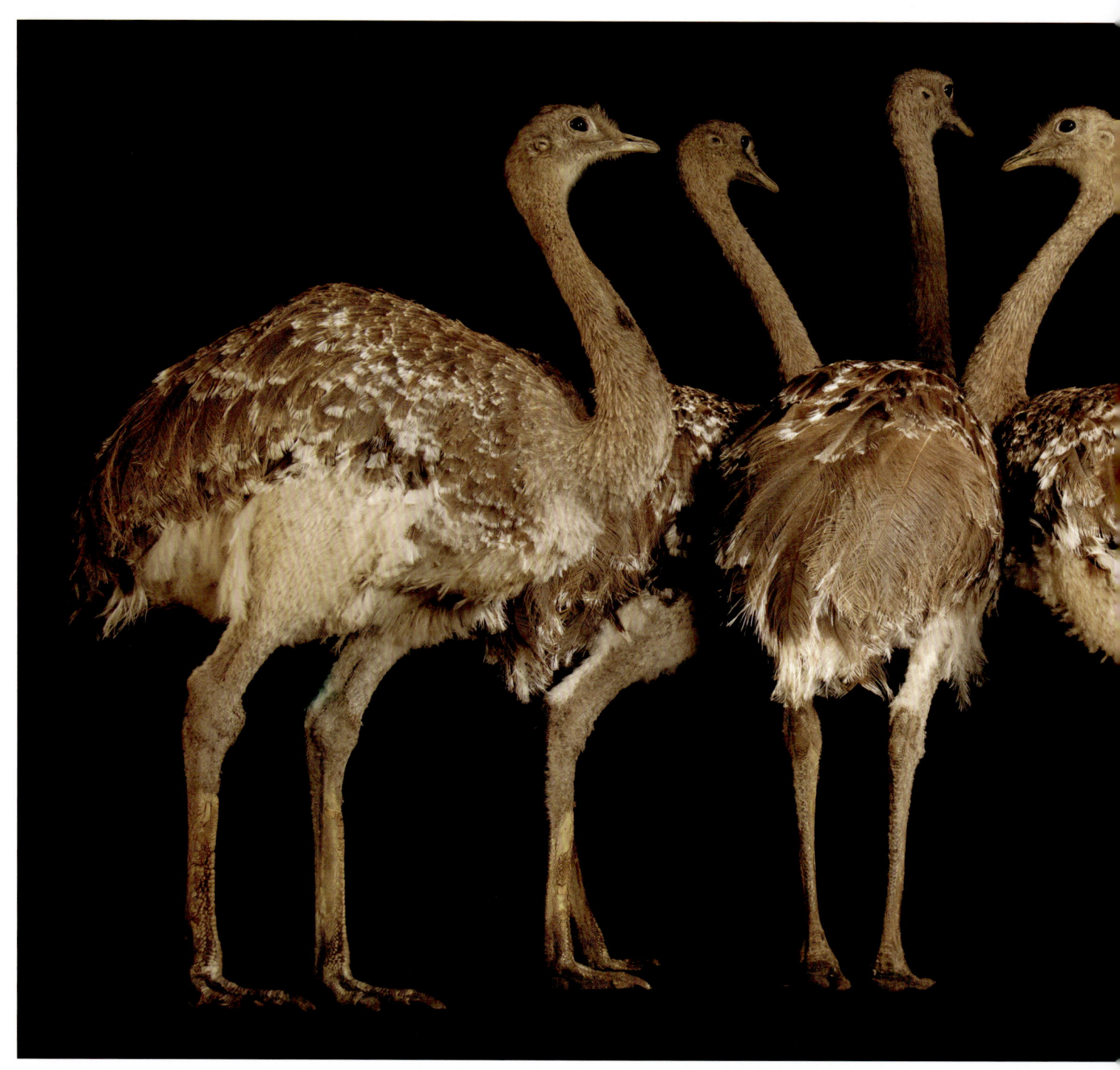

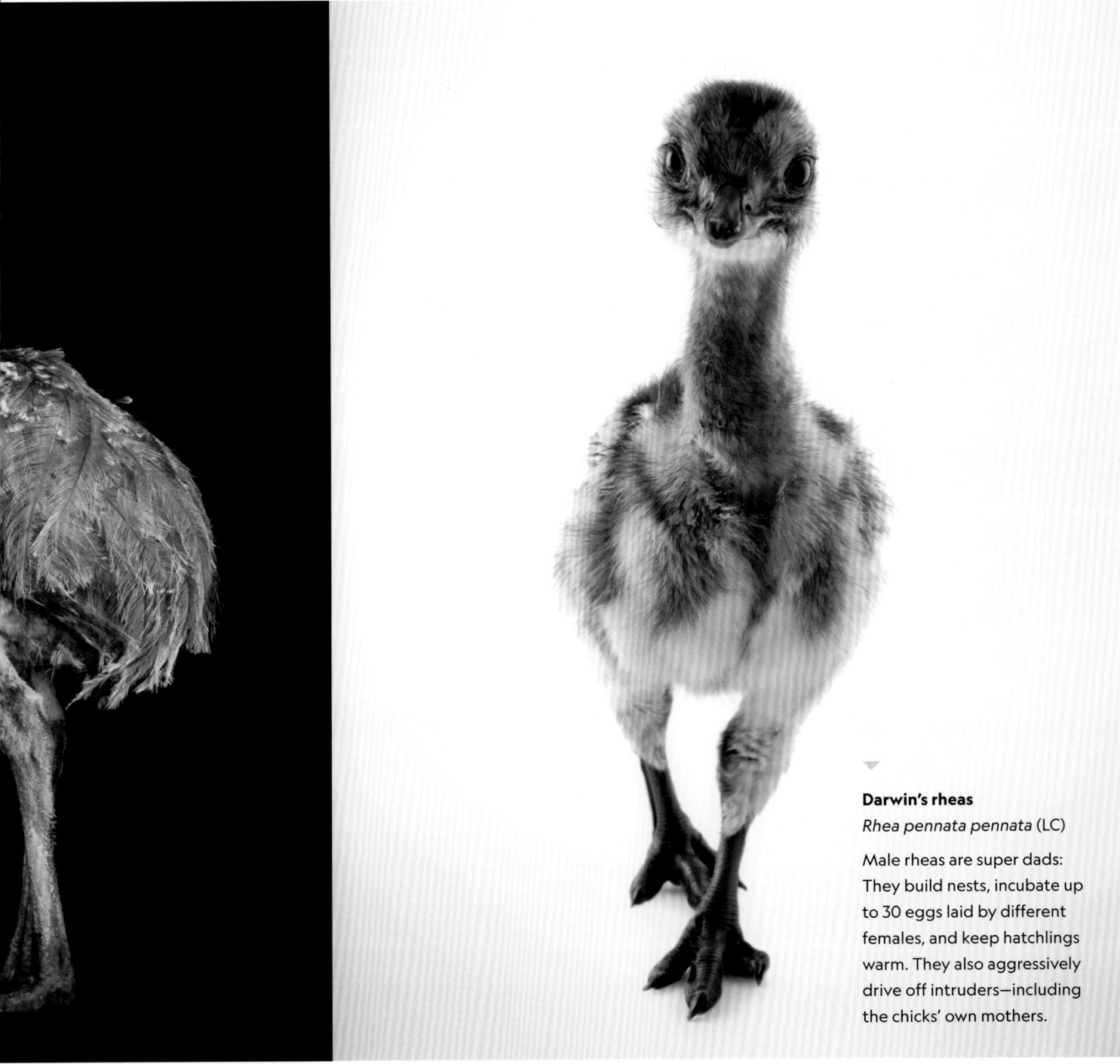

Darwin's rheas
Rhea pennata pennata (LC)

Male rheas are super dads:
They build nests, incubate up
to 30 eggs laid by different
females, and keep hatchlings
warm. They also aggressively
drive off intruders—including
the chicks' own mothers.

Virginia opossums
Didelphis virginiana virginiana (LC)

After a short pregnancy, opossum mothers give birth to bee-size babies. They crawl into Mom's pouch, in which they grow and nurse for up to 10 weeks before emerging and often hitching a ride on their mother as she roams the forest.

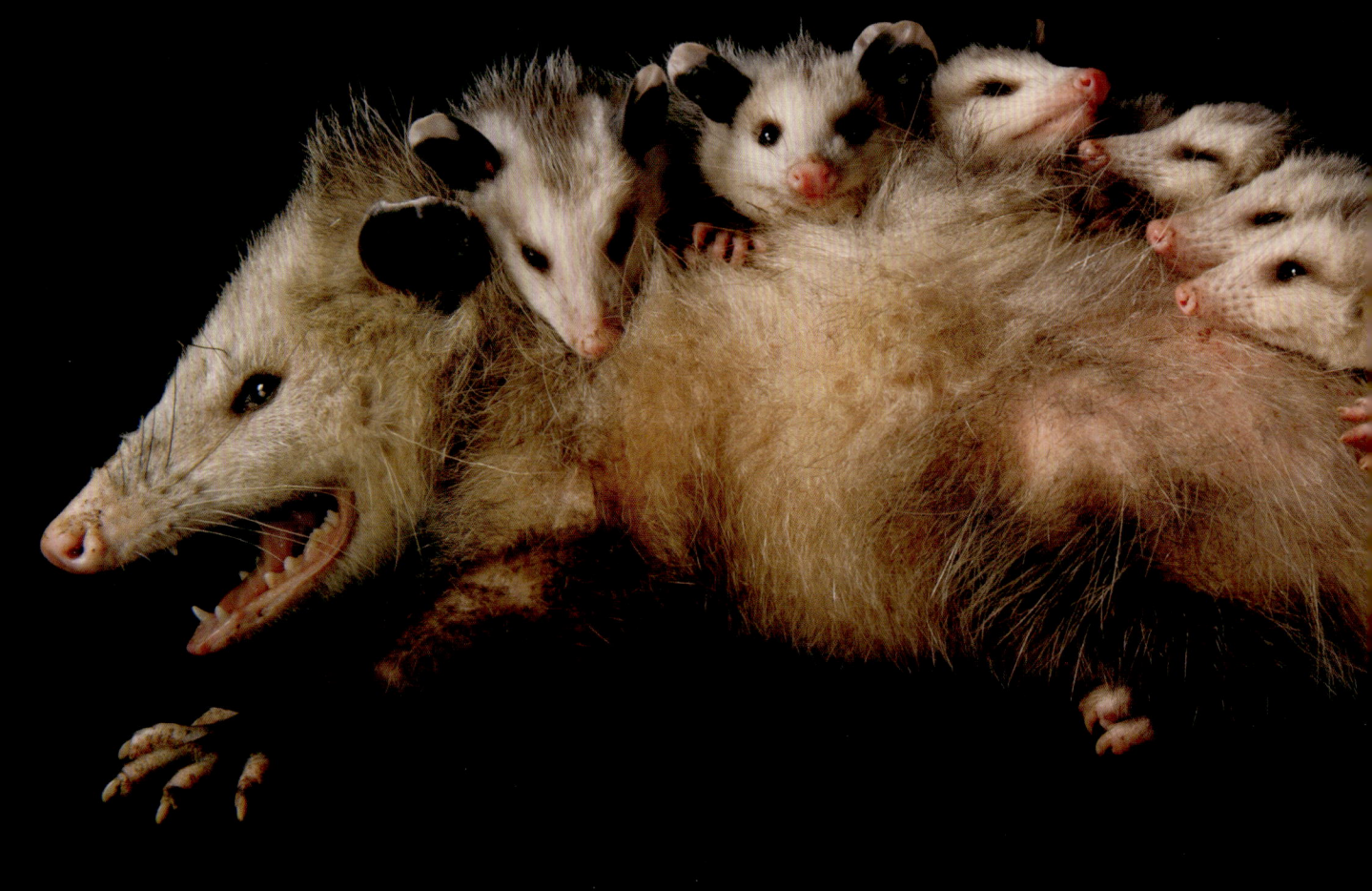

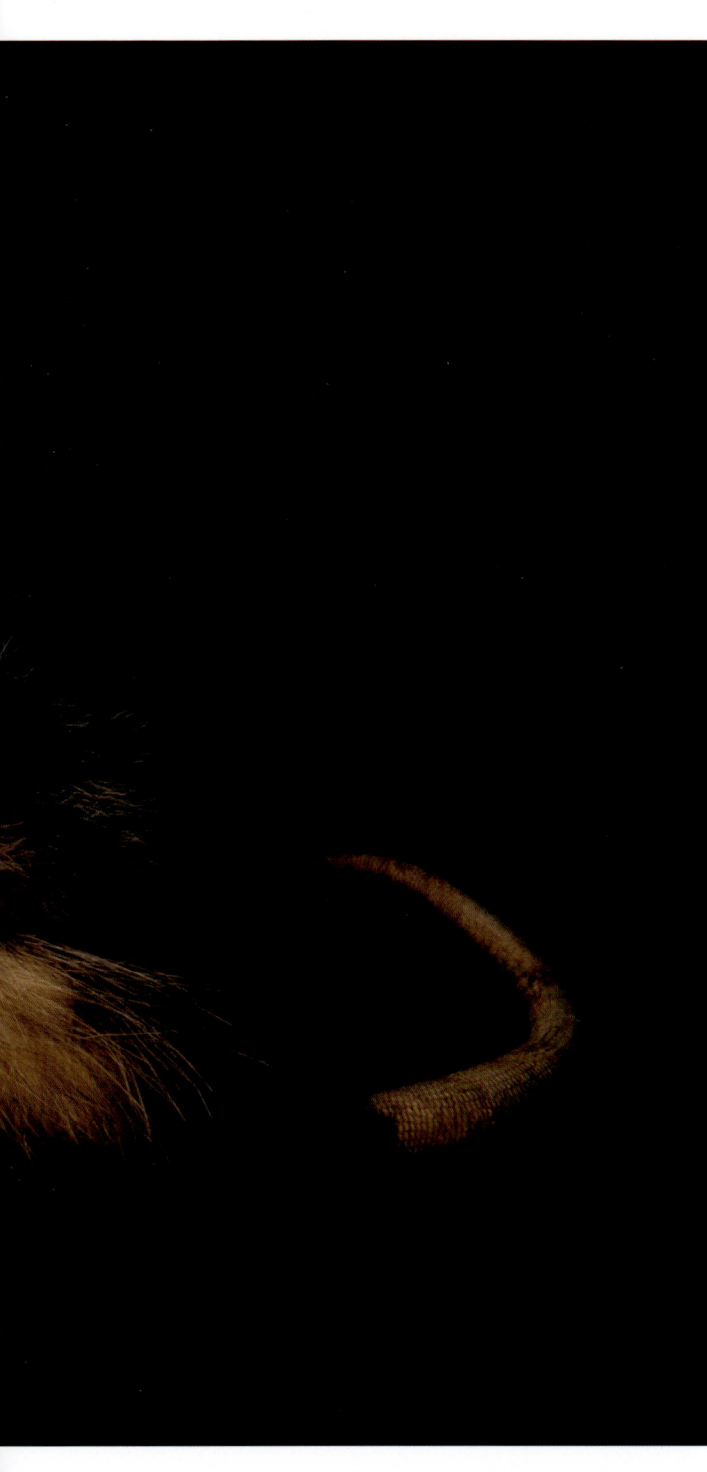

FAMILY PORTRAIT

ANIMAL: Virginia opossum, *Didelphis virginiana virginiana* (LC)

LOCATION: Nebraska Wildlife Rehab, Omaha, Nebraska

"My friend Laura Stastny from Nebraska Wildlife Rehab called me and said, 'I have an opossum; do you want to photograph it?' I said yeah. So she brings over a big pet kennel, pulls the lid off, and reaches in with gloved hands. I was expecting one possum, but she pulled out a mother possum with babies all over her back. Laura apologized, but I said, 'No, it's great. It looks better that way.' But I didn't want to go anywhere near Mom's head. She was snarling at me. She could bite me with big teeth. I was surprised the babies didn't fall off her back. They were all facing me in a line. All those babies stayed right there, looking at me. It was awesome."

BEACH
Rules

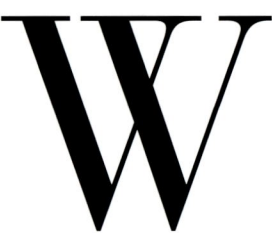

WITH SPINDLY LEGS and fluffy bodies, plover chicks are born to run. Although they weigh less than a scoop of ice cream when they're brand-new, these tiny chicks can walk and find food alongside their parents within hours of hatching. A few weeks later, fattened up, some young birds will leave their breeding grounds to join the fall migration. But to survive until migration, they need considerable help from their parents. Plover parents don't build nests. They lay their camouflaged eggs in a scrape of sand or gravel near beaches; the mottled eggs look like pebbles. Parents use a variety of tricks to draw predators' attention away from the eggs and, later, their chicks: They build fake nests, pretend to incubate nonexistent eggs, and feign injury, all to distract interlopers and protect their puffy, sand-colored youngsters.

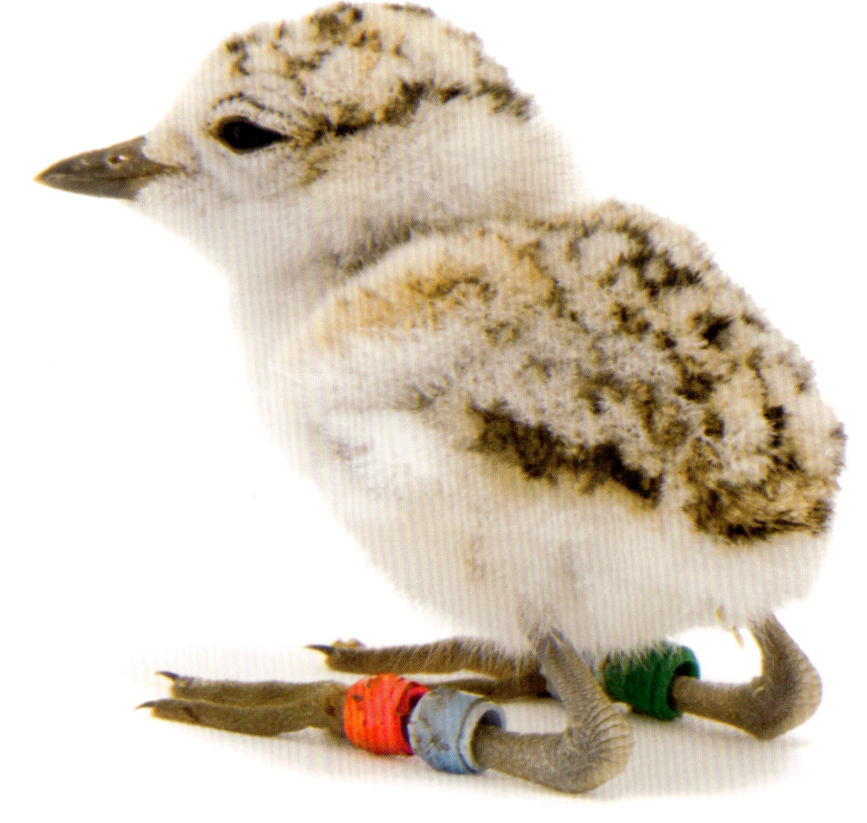

Western snowy plovers
Charadrius nivosus nivosus (NT)

Plover chicks leave their nest just a few hours after hatching. When a parent signals that a threat is near, the little ones flatten themselves on the sand, becoming almost invisible.

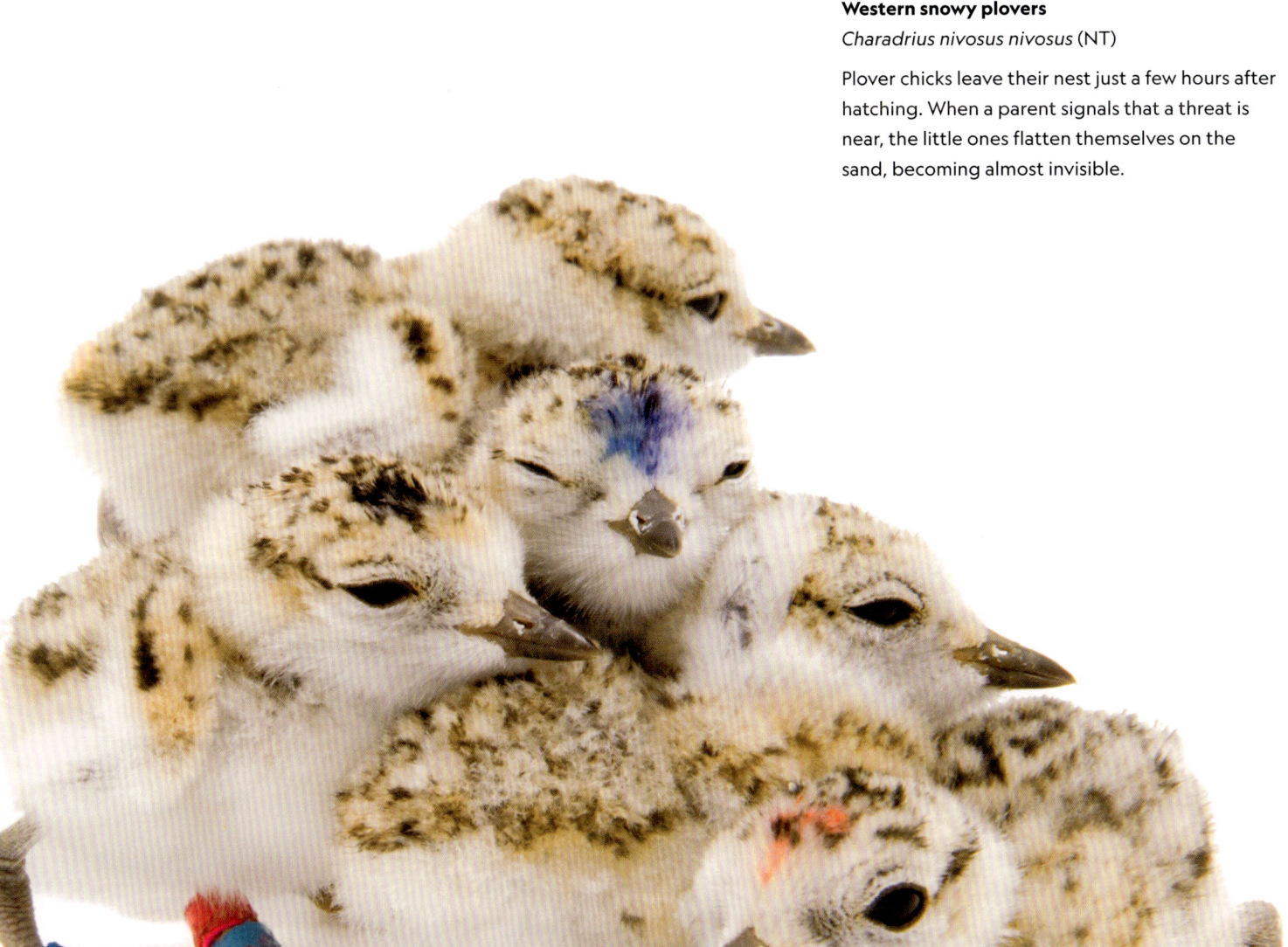

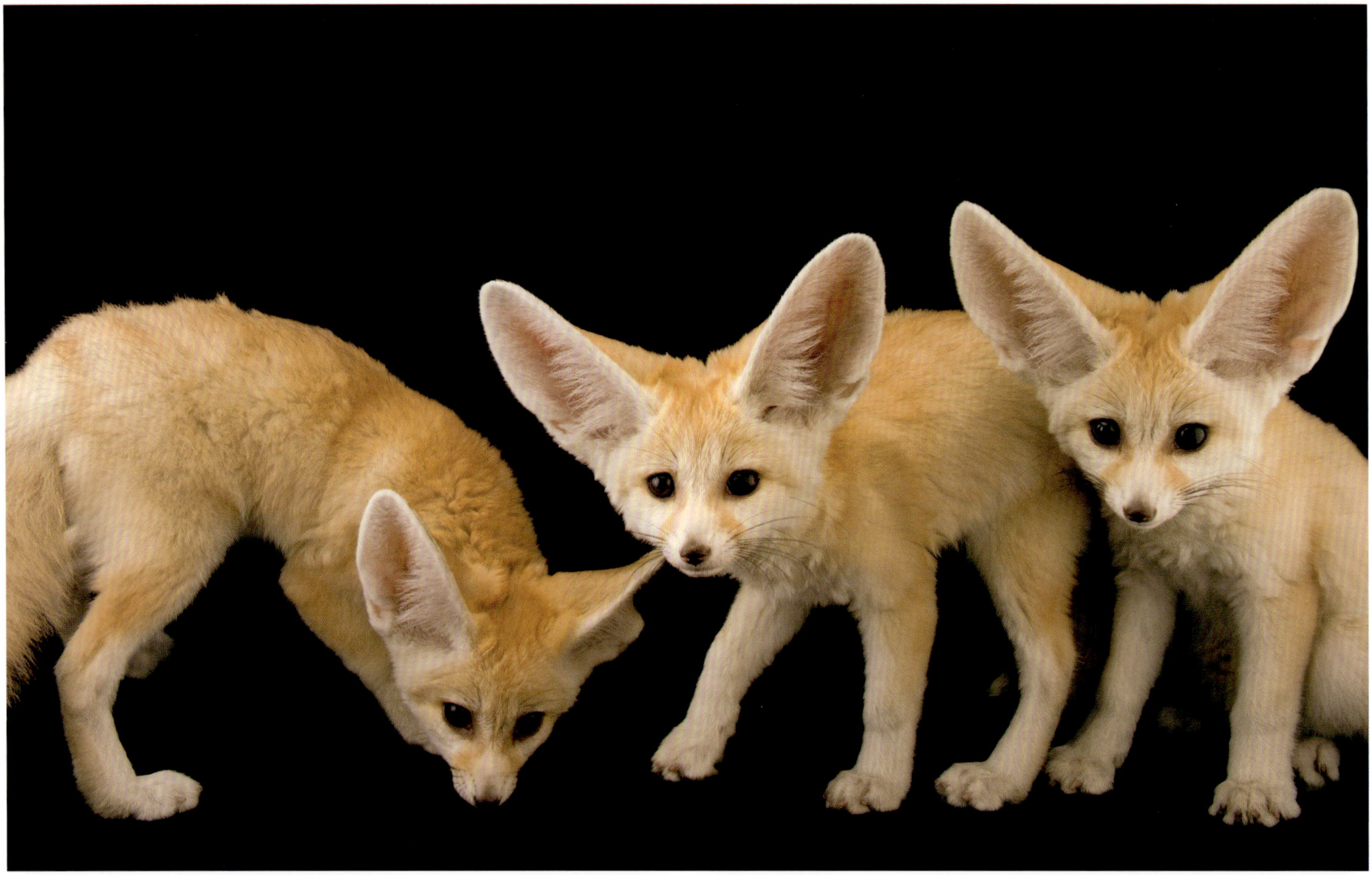

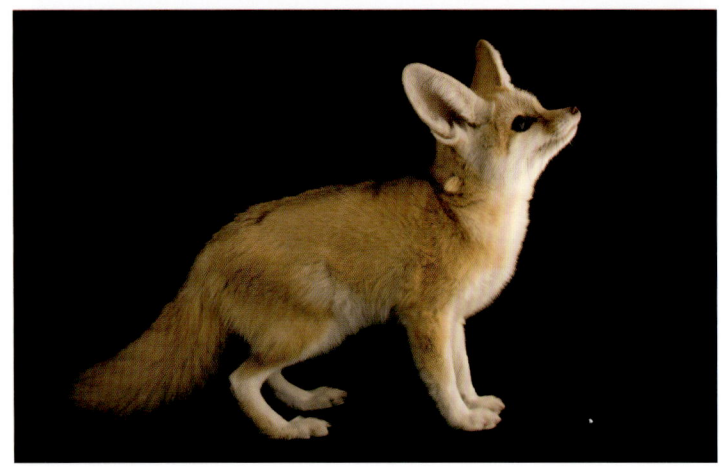
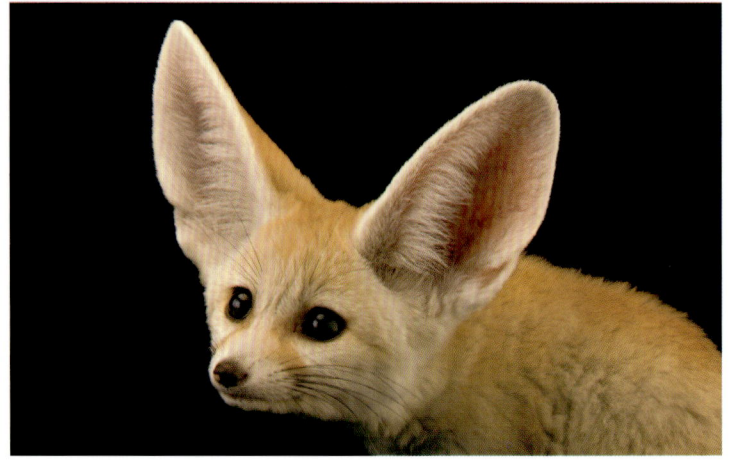

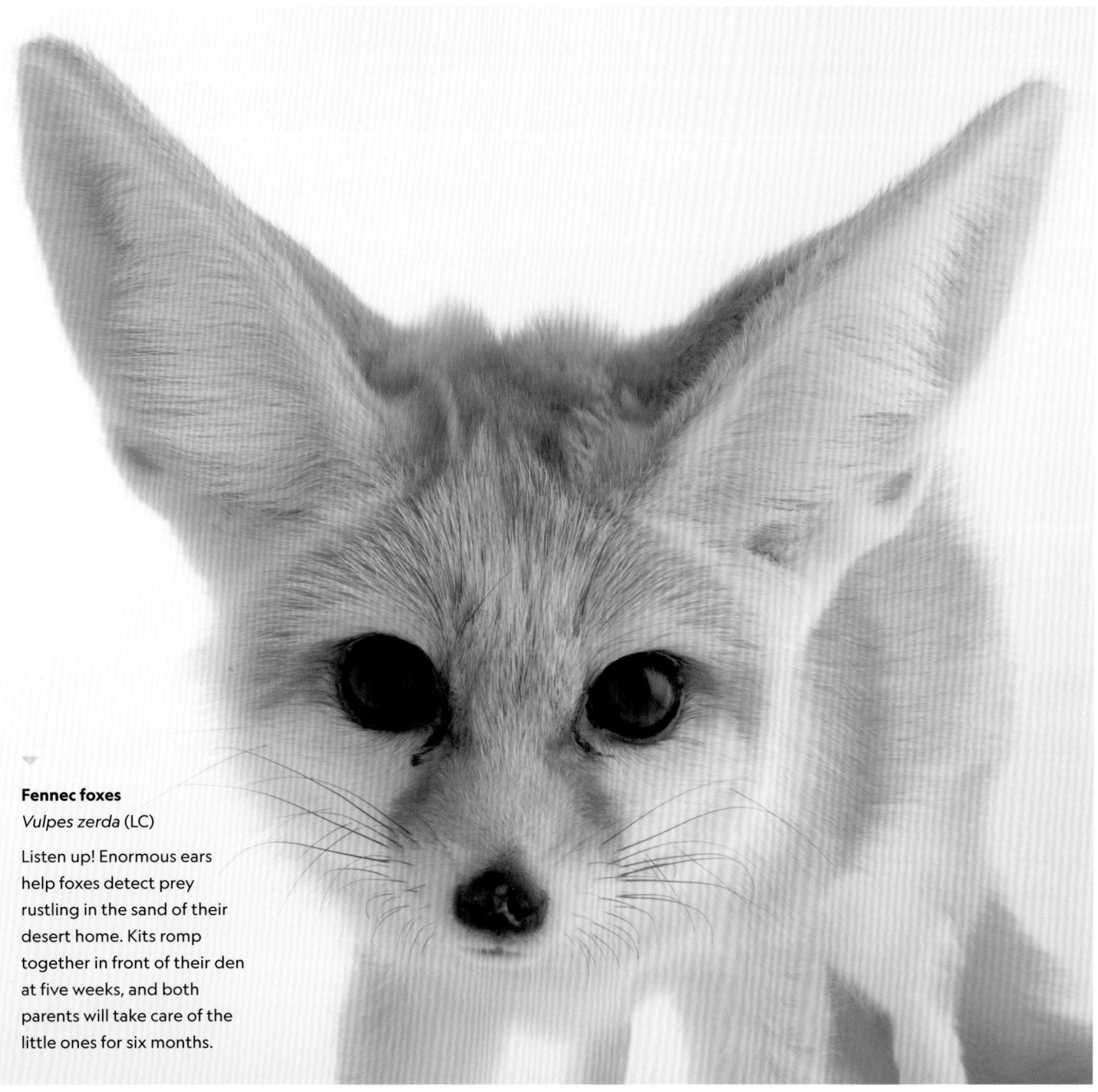

Fennec foxes

Vulpes zerda (LC)

Listen up! Enormous ears help foxes detect prey rustling in the sand of their desert home. Kits romp together in front of their den at five weeks, and both parents will take care of the little ones for six months.

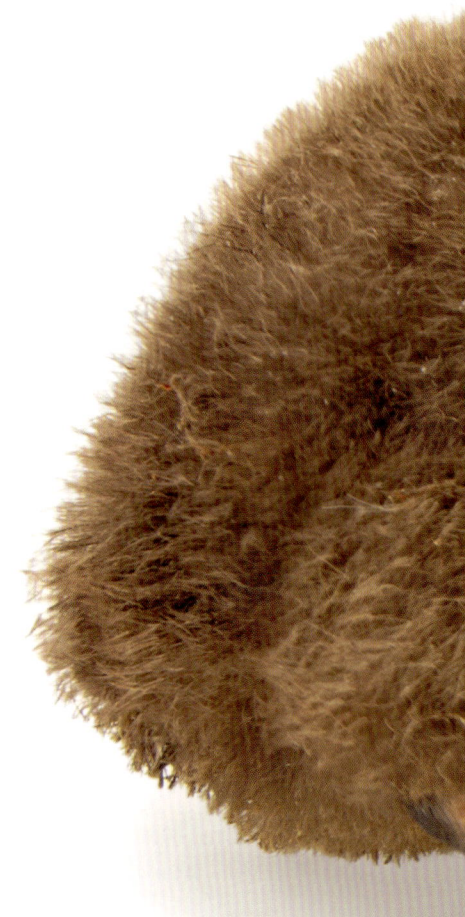

"

"We keep endangered species in breeding programs, hoping things will get better. But we also need to make sure a habitat is there for when the babies grow up."

▼

Little blue penguins

Eudyptula minor novaehollandiae (LC)

Parents keep their chicks close and warm for several weeks after hatching, but then watch them only at night until the chicks fledge. The youngsters are fully independent at around 10 weeks.

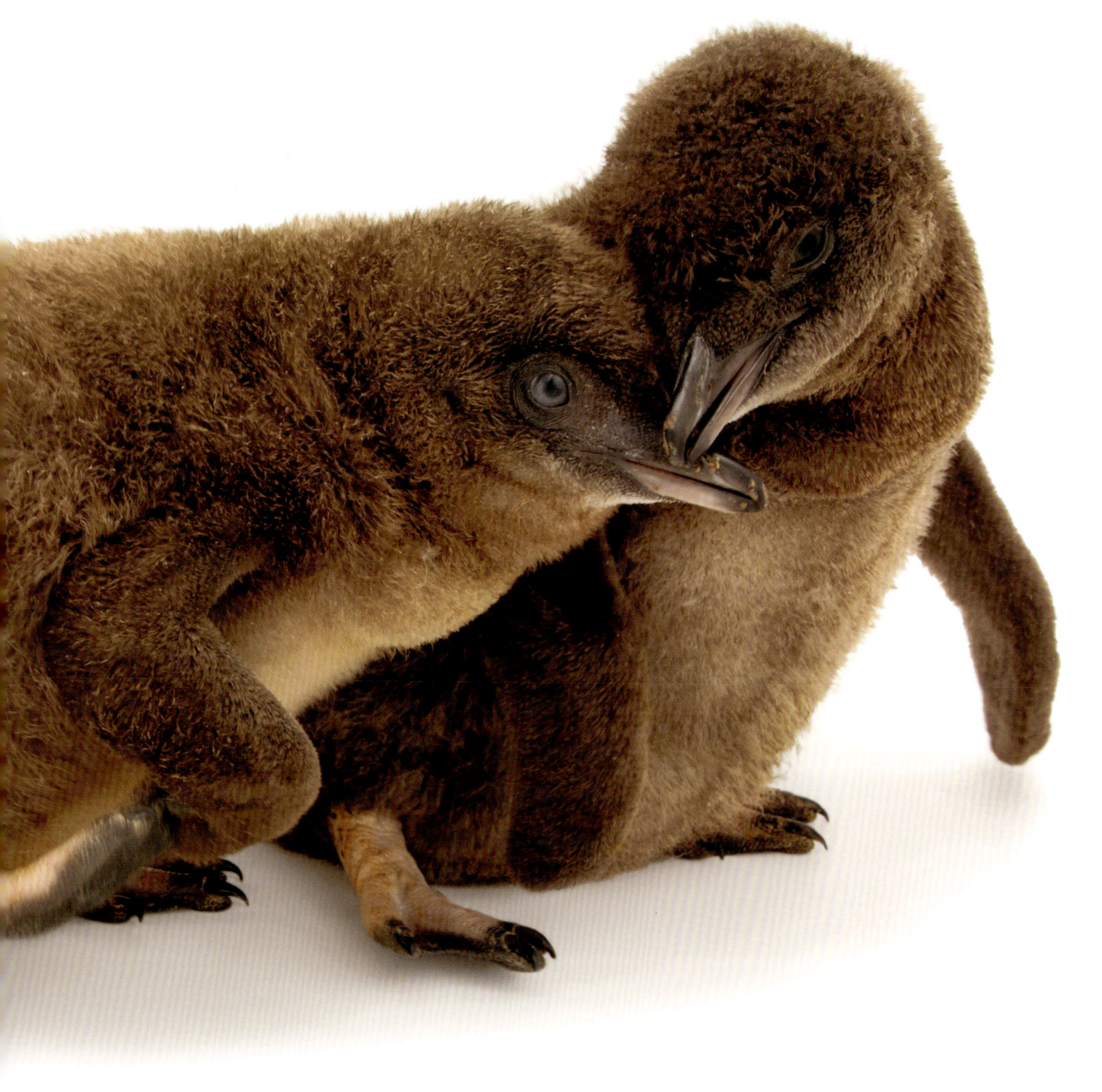

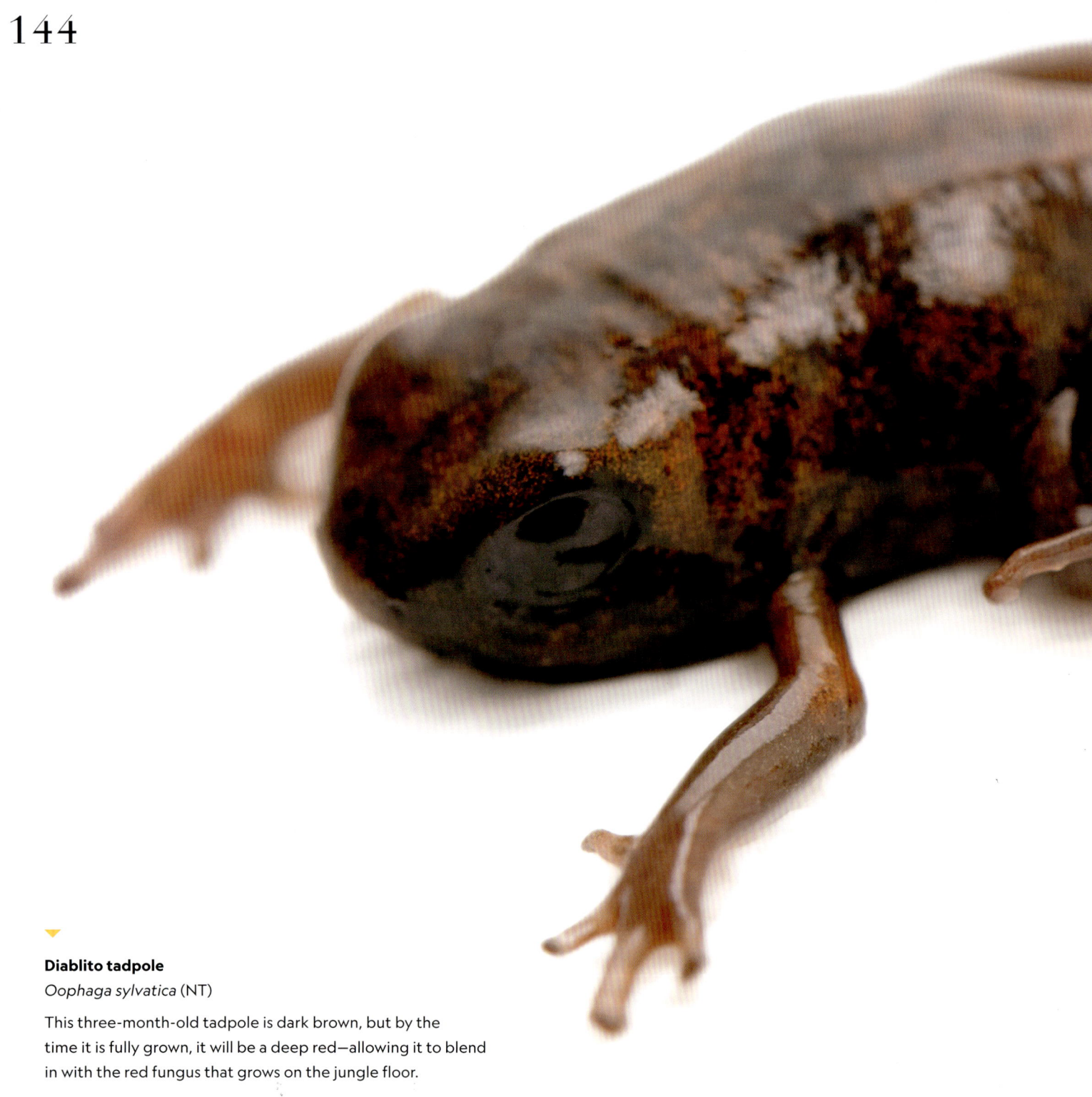

Diablito tadpole

Oophaga sylvatica (NT)

This three-month-old tadpole is dark brown, but by the time it is fully grown, it will be a deep red—allowing it to blend in with the red fungus that grows on the jungle floor.

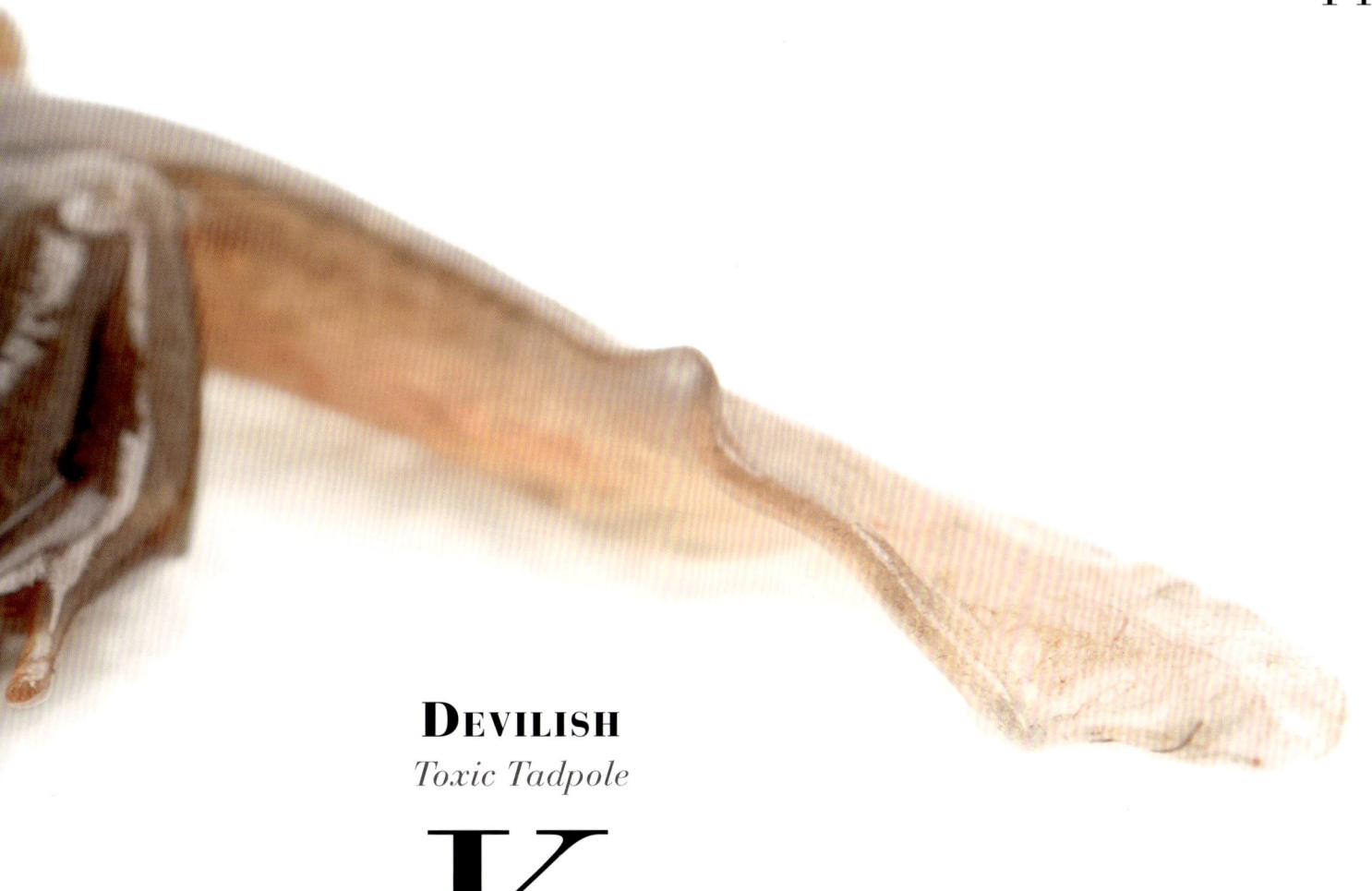

DEVILISH
Toxic Tadpole

K

KNOWN AS THE "little devil" in Spanish, this tiny poison frog lives in the lowland rainforests of Colombia and Ecuador. After the mother diablito's eggs hatch, she carries her tadpoles one by one on her back to a new spot. She leaves the tadpoles in a pool of rainwater collected in a bromeliad plant's axil, a little reservoir where the leaves of the plant meet the stem. She checks on the tadpoles regularly and lays unfertilized eggs for them to eat—the tadpoles' only food until their metamorphosis into frogs.

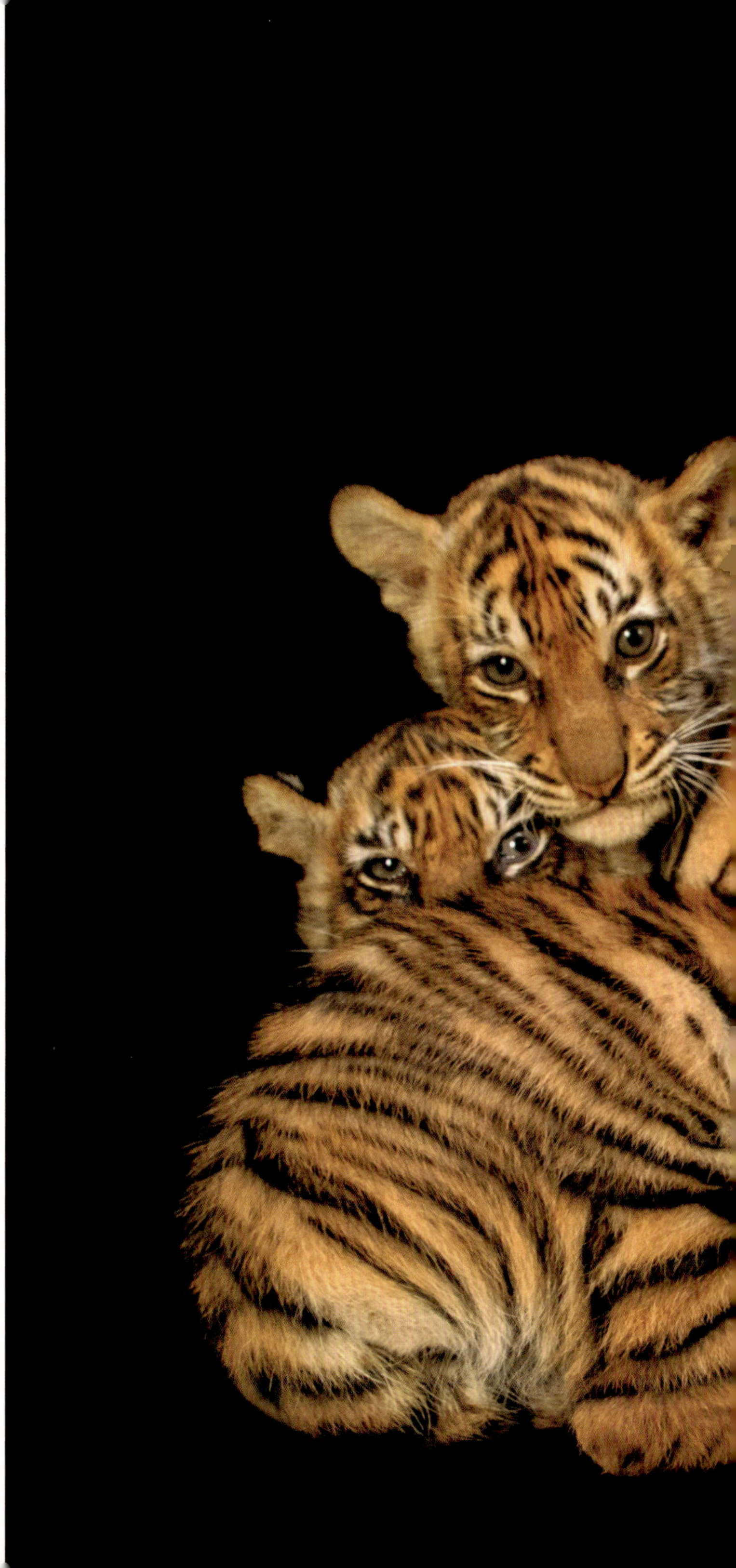

Sumatran tigers

Panthera tigris sumatrae (CR)

Tiger families are close-knit. Cubs nurse for three months, then by five or six months they accompany their mom as she hunts. Some tigers stay with their mother for three years.

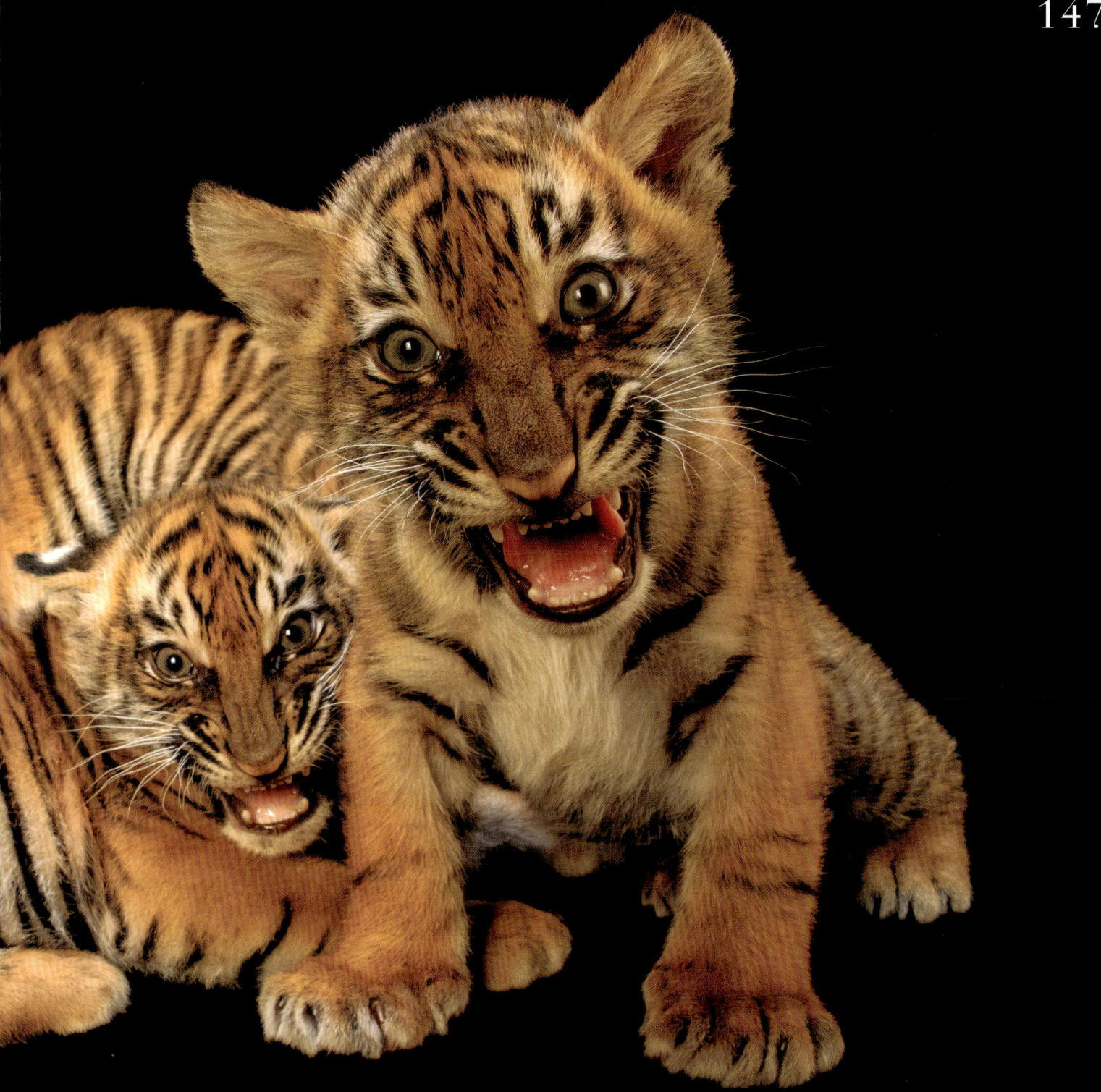

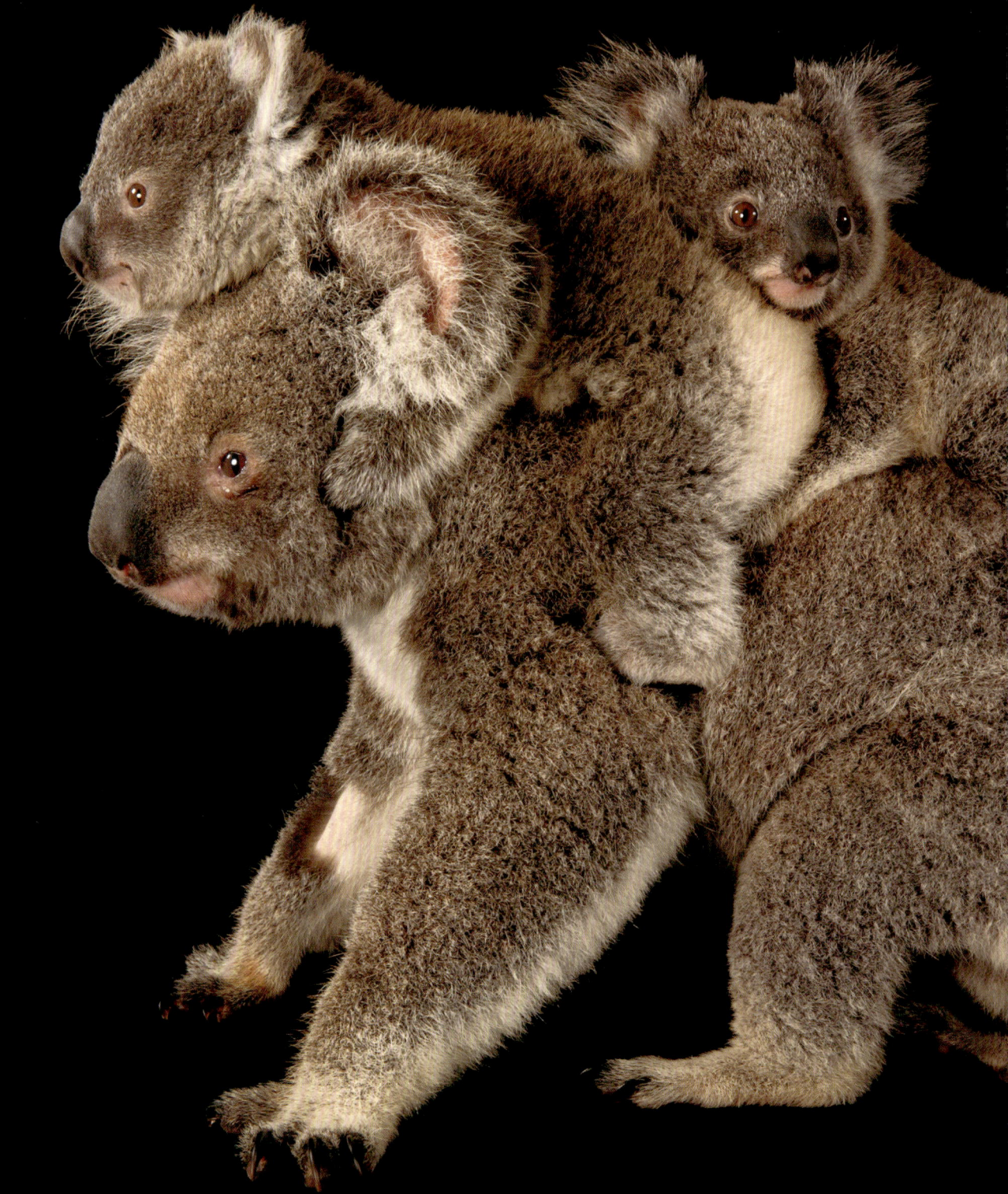

> "Koala moms are famous
> for readily taking on
> a second child that's not
> their own. This mother gave
> birth to one joey and
> adopted the other."

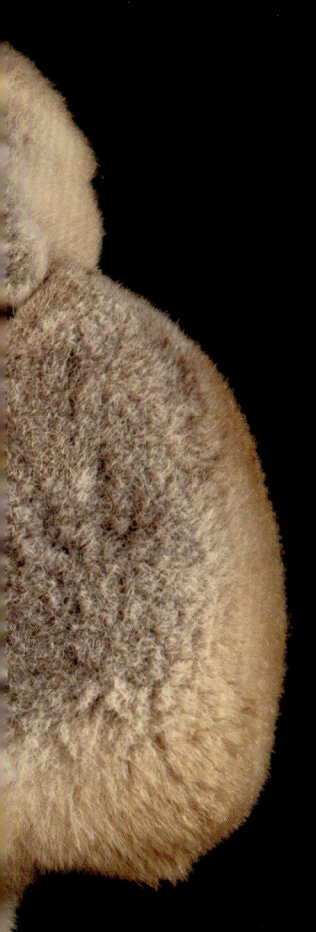

Koalas
Phascolarctos cinereus (VU)

A newborn koala, called a joey, resembles a pink jelly bean. It stays in its mother's pouch for seven months, nursing or eating predigested leaves, and remains with her until the next joey is born.

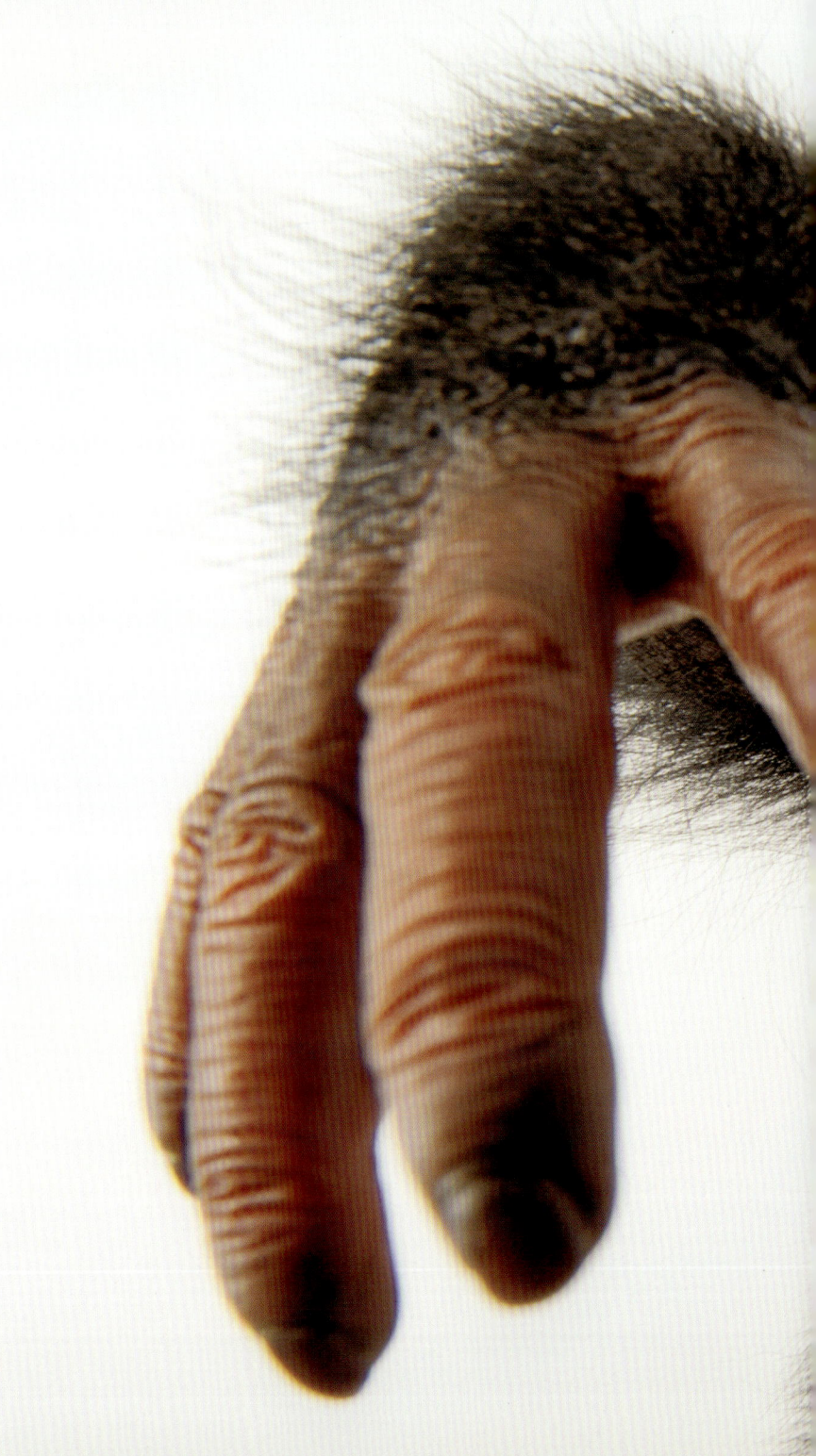

Chimpanzee

Pan troglodytes (EN)

Smart and gregarious, chimpanzees form strong relationships with siblings and friends. Bonds with juvenile playmates can figure prominently later in the complex social hierarchy of adult chimps.

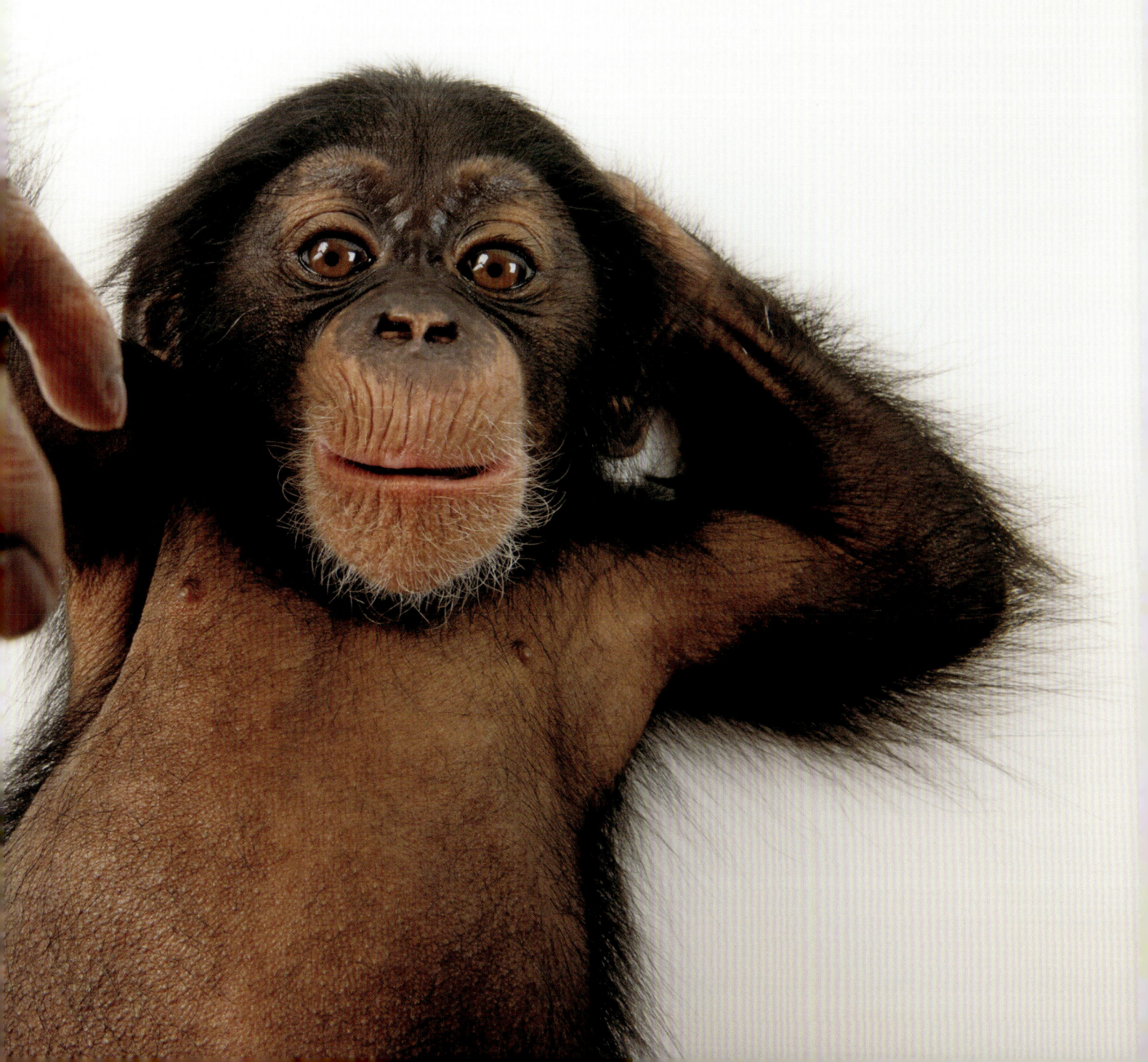

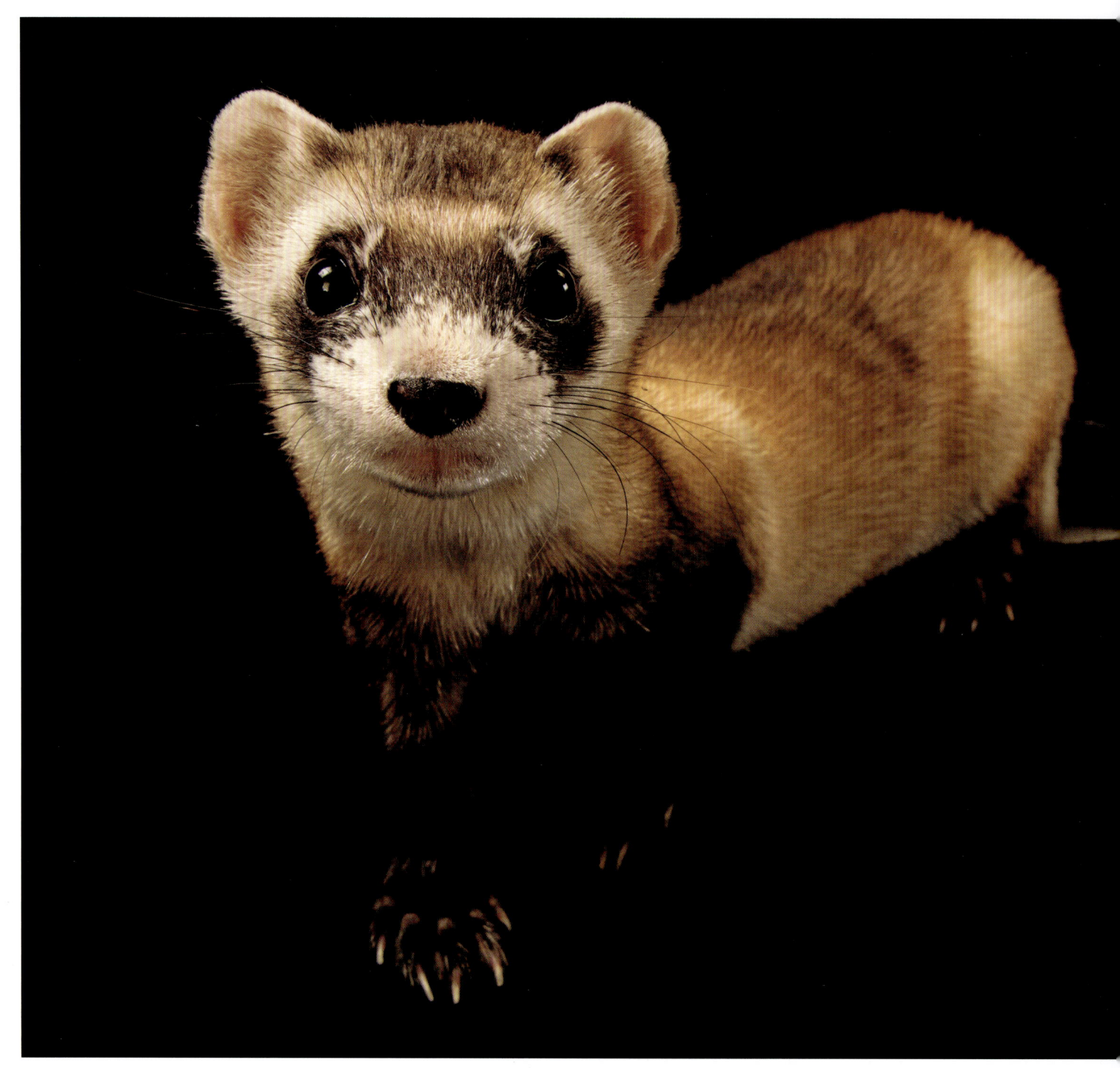

Black-footed ferret

Mustela nigripes (EN)

Ferrets are solitary animals. Their long, slender bodies enable them to crawl in and out of the burrows and dwellings of their primary prey, the prairie dog.

OUTDOOR EDUCATION

The black-footed ferret is one of the world's rarest mammals and the only ferret species native to North America. The species was thought to be extinct until a small population was discovered in Wyoming in 1981. Through a combination of scientific advancements, including artificial insemination and genetic cloning, conservationists have used that last wild population to bring the species back from the brink. It's considered one of the most impressive feats of conservation ever attempted.

But scientists have had to figure out more than just the technicalities of breeding ferrets in human care. If captive-bred ferrets are released into the wild without learning how to dodge raptors, navigate a burrow, take cover in the cold, or hunt a prairie dog, they won't last long. So how can people teach ferrets the survival skills they would otherwise learn from their mothers?

At the National Black-Footed Ferret Conservation Center in Colorado, researchers place ferrets destined for reintroduction to the wild in a "preconditioning" program. Researchers give ferrets their first taste—literally—of prairie dogs; they use stuffed raptors to swoop down on the ferrets; and they have even created a "robo-badger" to simulate a typical predator. The program helps the little ones stay safe, find food, and learn their way around the neighborhood—a big lift by humans to simulate what Mom does best.

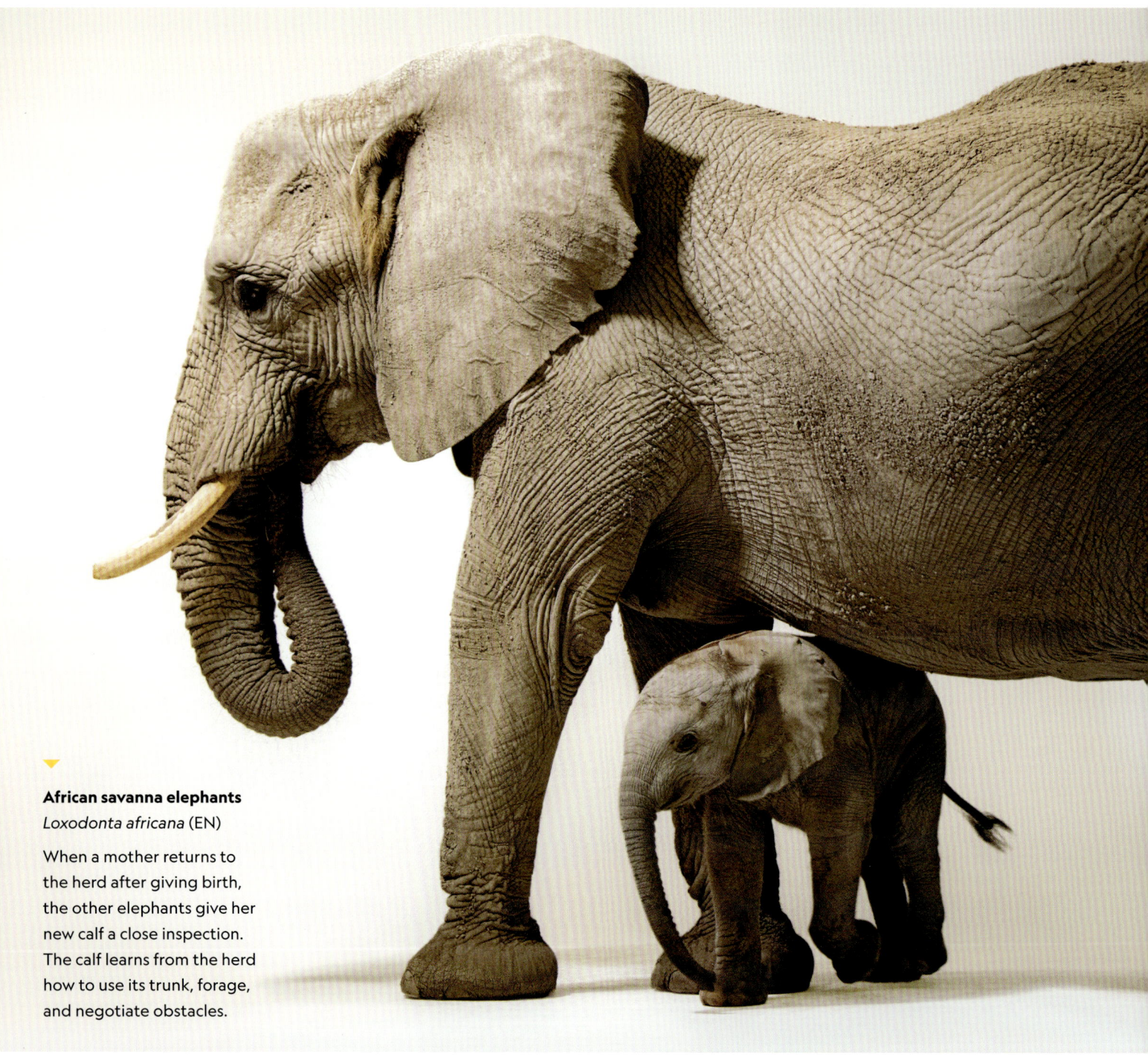

African savanna elephants

Loxodonta africana (EN)

When a mother returns to the herd after giving birth, the other elephants give her new calf a close inspection. The calf learns from the herd how to use its trunk, forage, and negotiate obstacles.

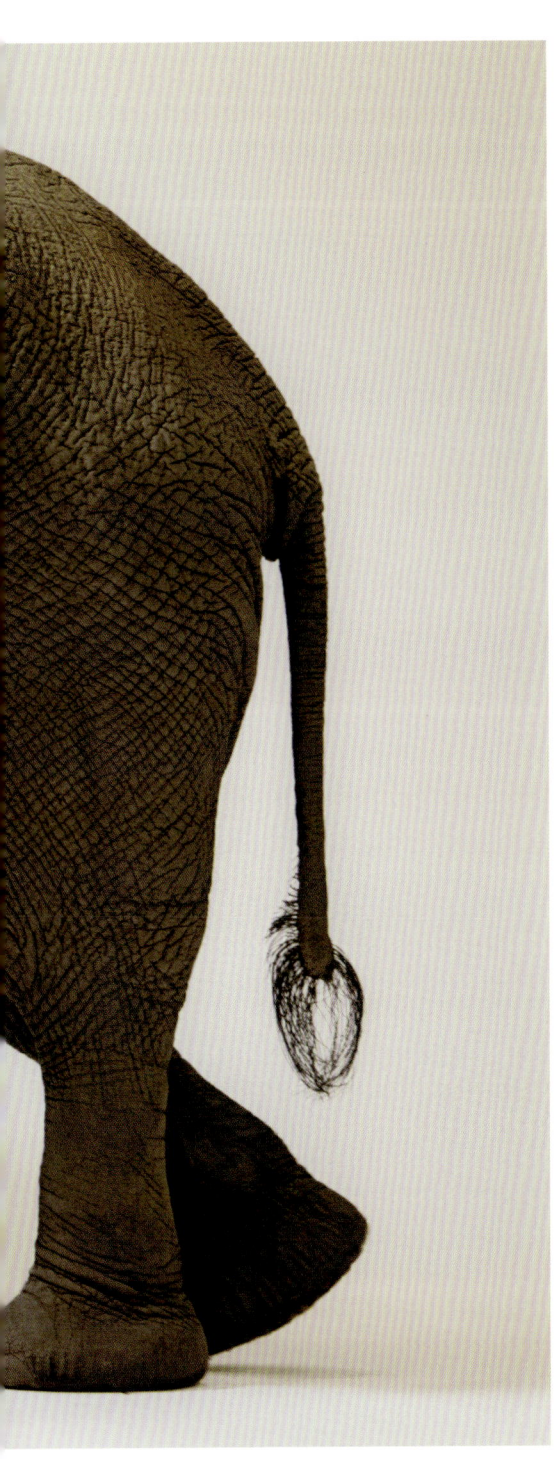

DOTING MOTHER
Years-long Care

I

IN THE MATRIARCHAL world of African savanna elephants, Mother knows best—and family ties are unbreakable. Elephant mothers have such a long pregnancy and are so devoted to their little ones that they give birth only once every two to four years. They usually have only four calves during their lifetime. A calf stays very close to its mother for its first few years of life, but Mom isn't the only one raising this big baby: The other elephants in the all-female herd—aunties, cousins, sisters—also help care for the calf. If the youngster is a female, she will spend her entire life with this herd (males go off on their own once they reach maturity). And if her mother is the matriarch? Someday she'll take on the same role.

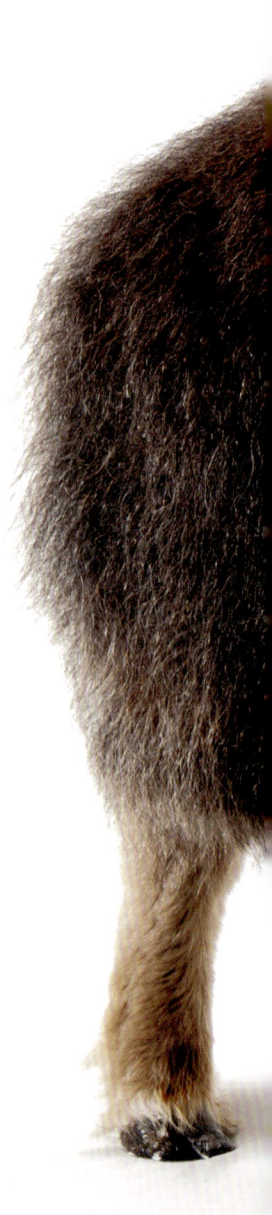

Muskox

Ovibos moschatus moschatus (LC)

To survive in the Arctic, muskox calves stay with
their mothers for two years. Cows provide warmth
until the calves grow their thick, insulating coat of fur;
the larger herd provides protection.

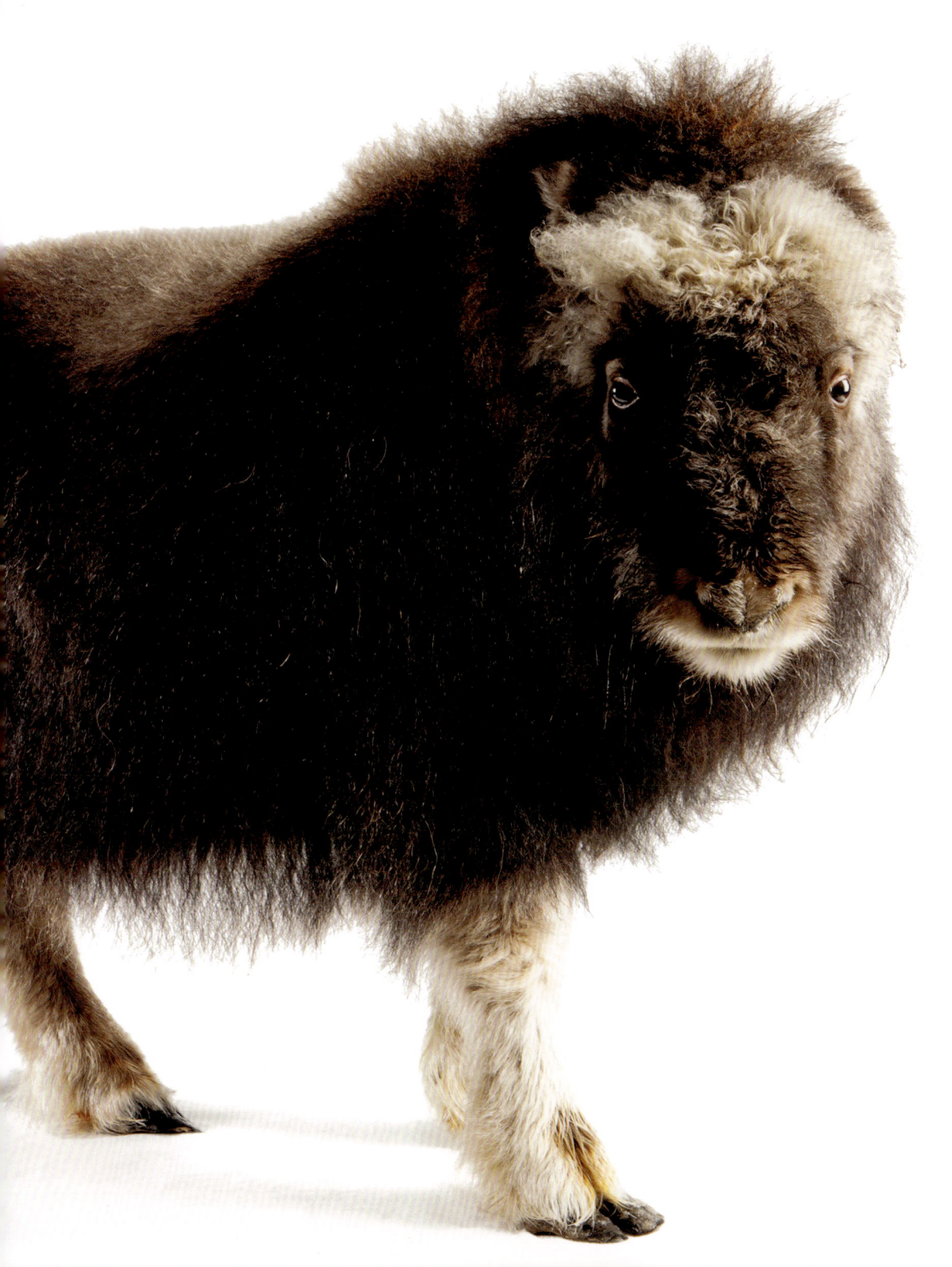

**Bornean orangutan (infant) and
Bornean/Sumatran cross (adult)**
Pongo pygmaeus (CR) and
Pongo pygmaeus x abelii

This infant orangutan hangs on its adoptive mother. Orangutans have the longest gap between births of any mammal—up to 10 years—because of how much attention they give while raising their young. Only after age three do orangutan babies start to move around and feed on their own.

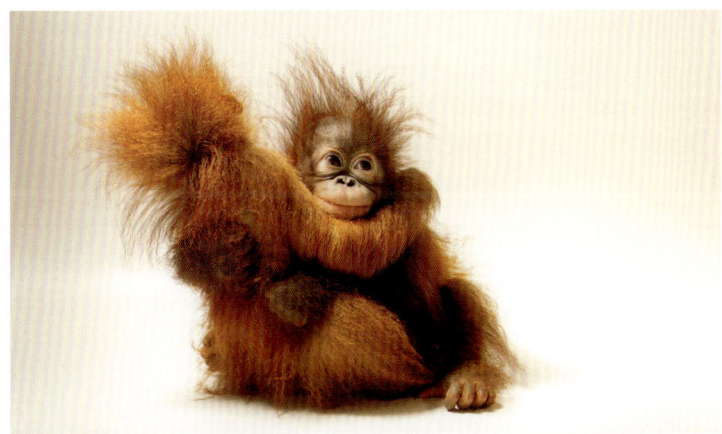

Sumatran orangutan *Pongo abelii* (CR, left) and **Bornean orangutan** *Pongo pygmaeus* (CR)

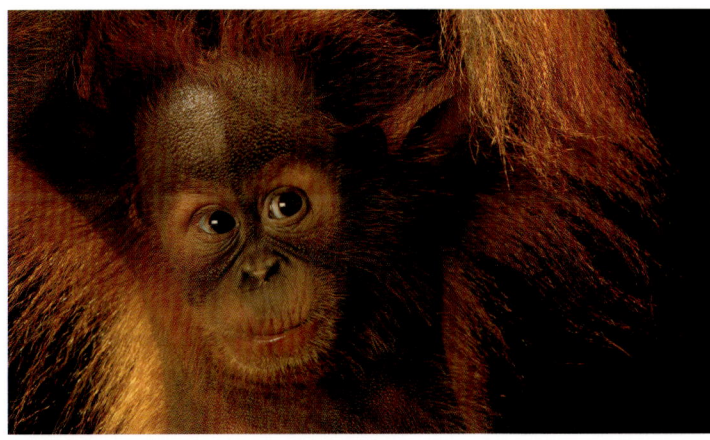

Sumatran orangutan *Pongo abelii* (CR)

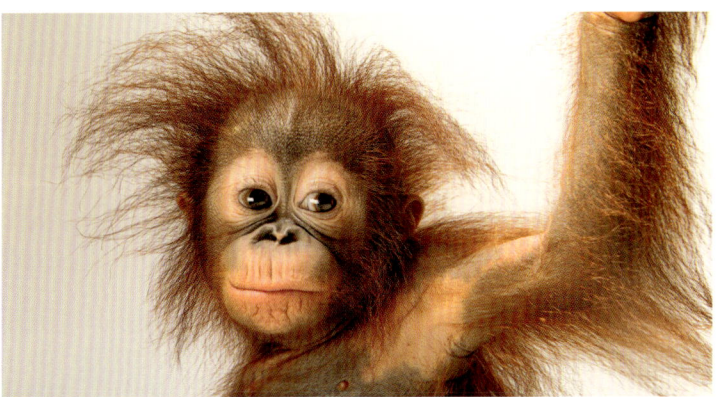

Bornean orangutan *Pongo pygmaeus* (CR)

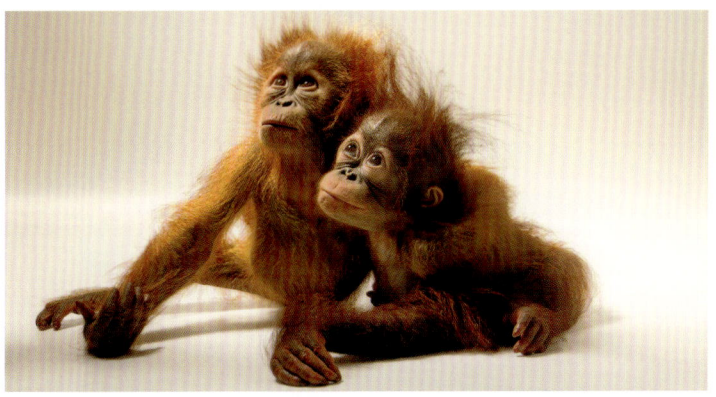

Sumatran orangutan *Pongo abelii* (CR, left) and **Bornean orangutan** *Pongo pygmaeus* (CR)

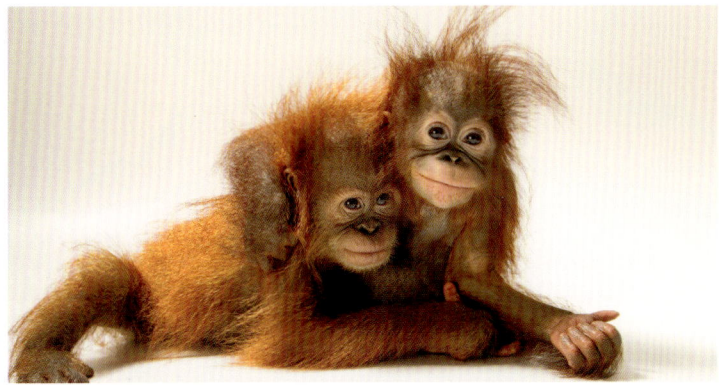

Sumatran orangutan *Pongo abelii* (CR, left) and **Bornean orangutan** *Pongo pygmaeus* (CR)

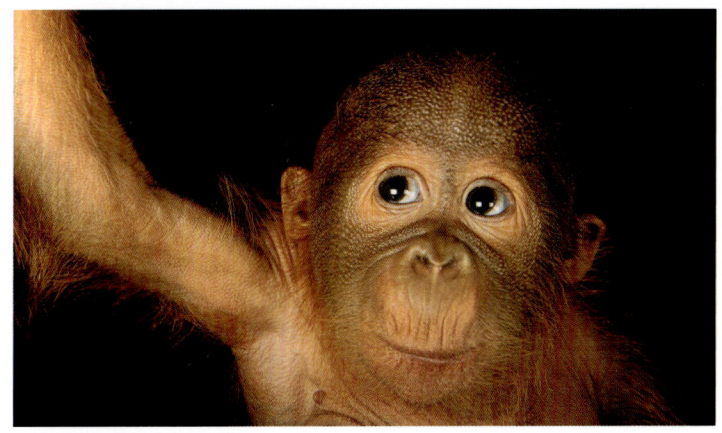

Bornean/Sumatran orangutan cross *Pongo pygmaeus x abelii*

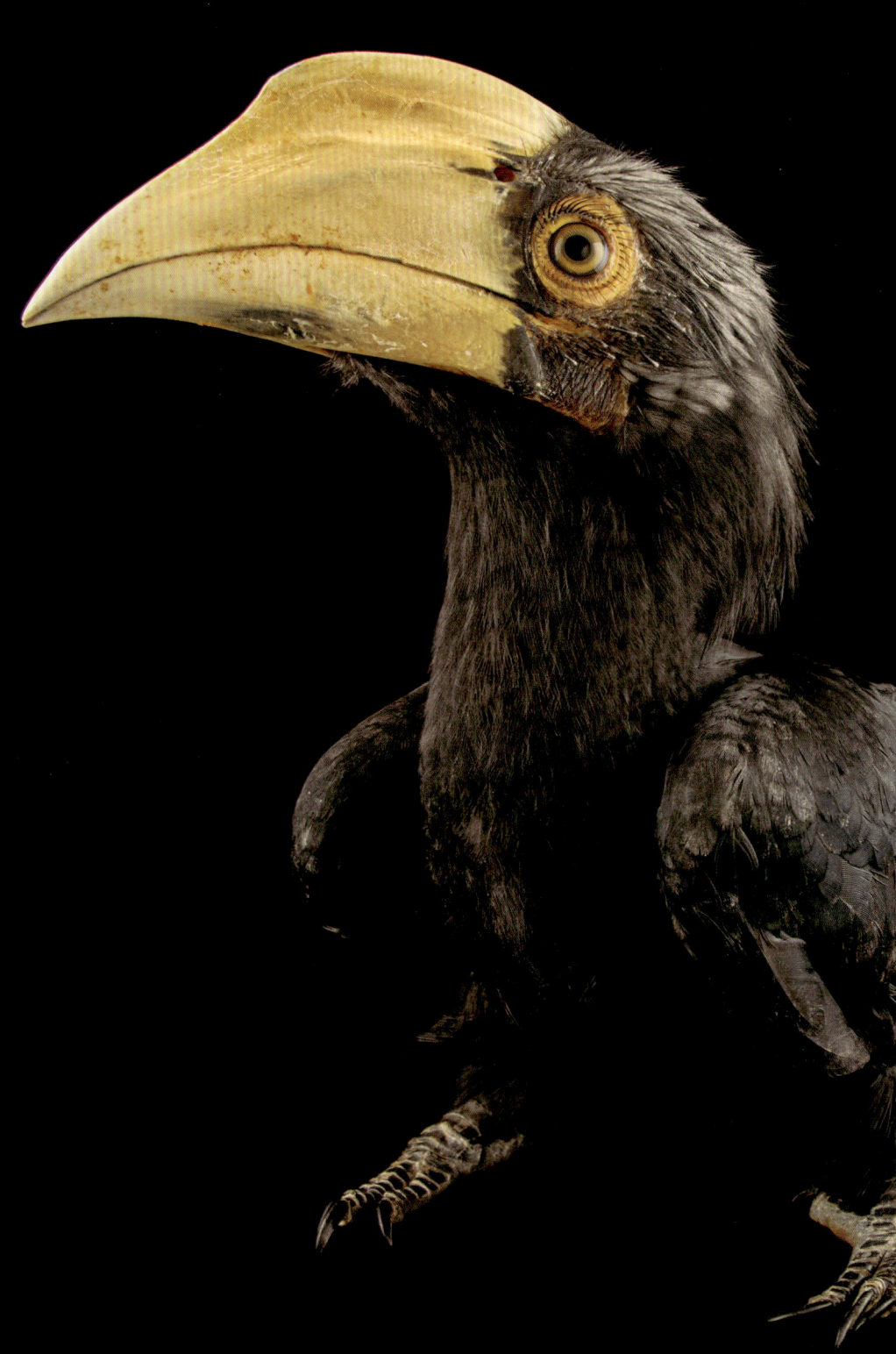

Black hornbill

Anthracoceros malayanus (VU)

This forest bird (pictured here as a juvenile) of Southeast Asia has an elaborate approach to parenting. When the chick finally fledges, it returns to stay with its parents for another six months.

WHERE'S MOM?
Expert Hider

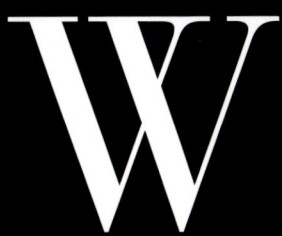

WHEN A FEMALE hornbill is ready to lay her eggs, she finds a suitable tree cavity and uses scraps of food and excrement to seal herself inside. Ensconced within her hiding spot, she lays the eggs and incubates them for 30 days. Meanwhile, her mate's job is to feed her: He forages for fruit and insects and returns to the tree, feeding the mother through a tiny opening. After the chicks hatch, the mother stays inside the tree for several weeks, caring for the youngsters. Finally she pecks her way out, leaving the young birds inside the hole after she seals it up again. When the fledgling hornbills are finally ready to fly, they will break open the nesting area one last time and fly out to join their parents.

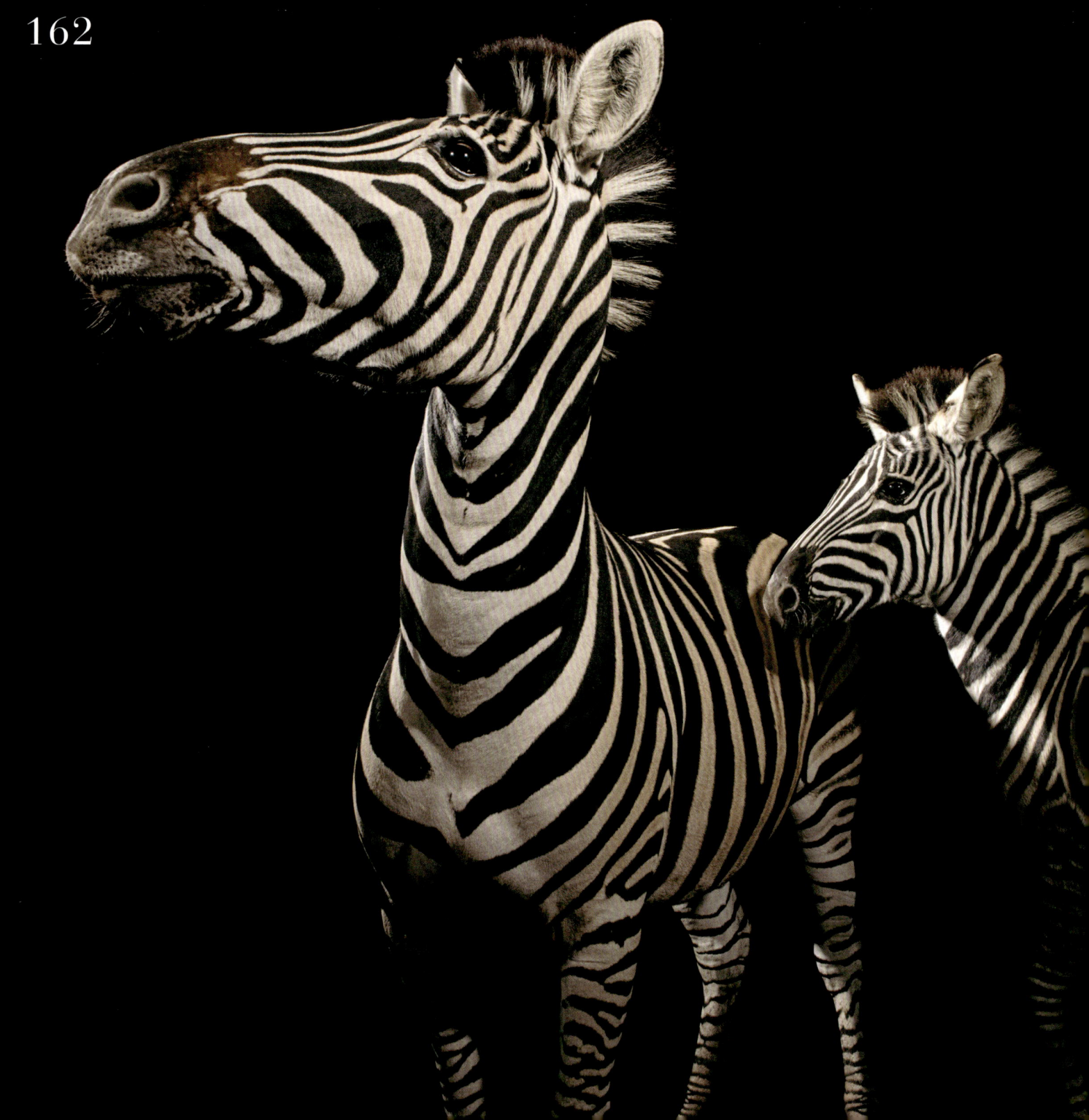

"For critically endangered species, there are not a ton of babies born. Often they won't give birth in captive breeding programs, and that's why it's so important to protect their habitats."

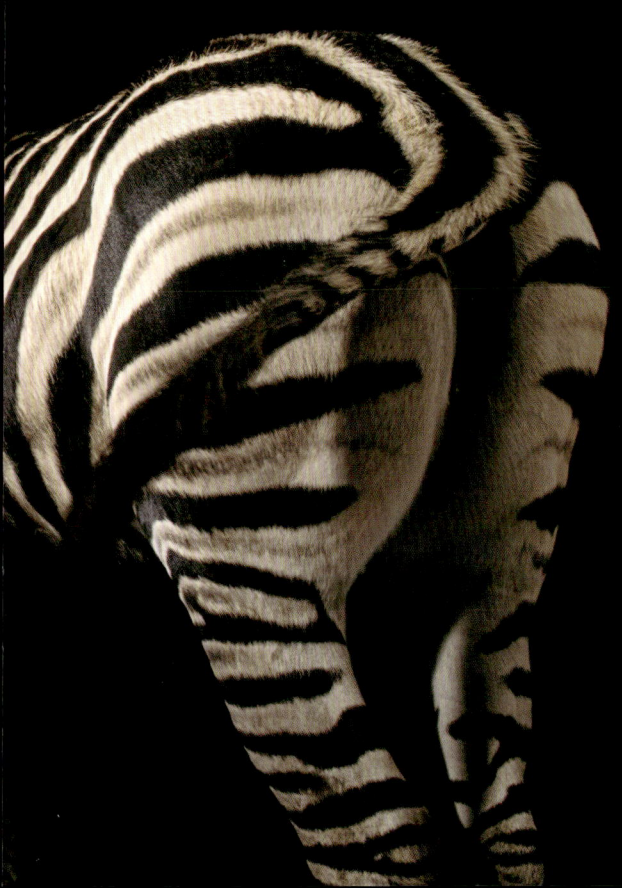

Chapman's zebras
Equus quagga chapmani (NT)

Zebras live in small family groups composed of a stallion, several mares, and their foals. If a hyena or wild dog approaches, the stallion may attack, while the rest of the family clusters together to protect their young.

Common marmosets

Callithrix jacchus (LC)

A mother marmoset embraces her three-month-old. Marmosets typically give birth to twins—a rarity among primates. Raising twins is a lot of work, so fathers, older siblings, and even nieces and nephews all lend a hand.

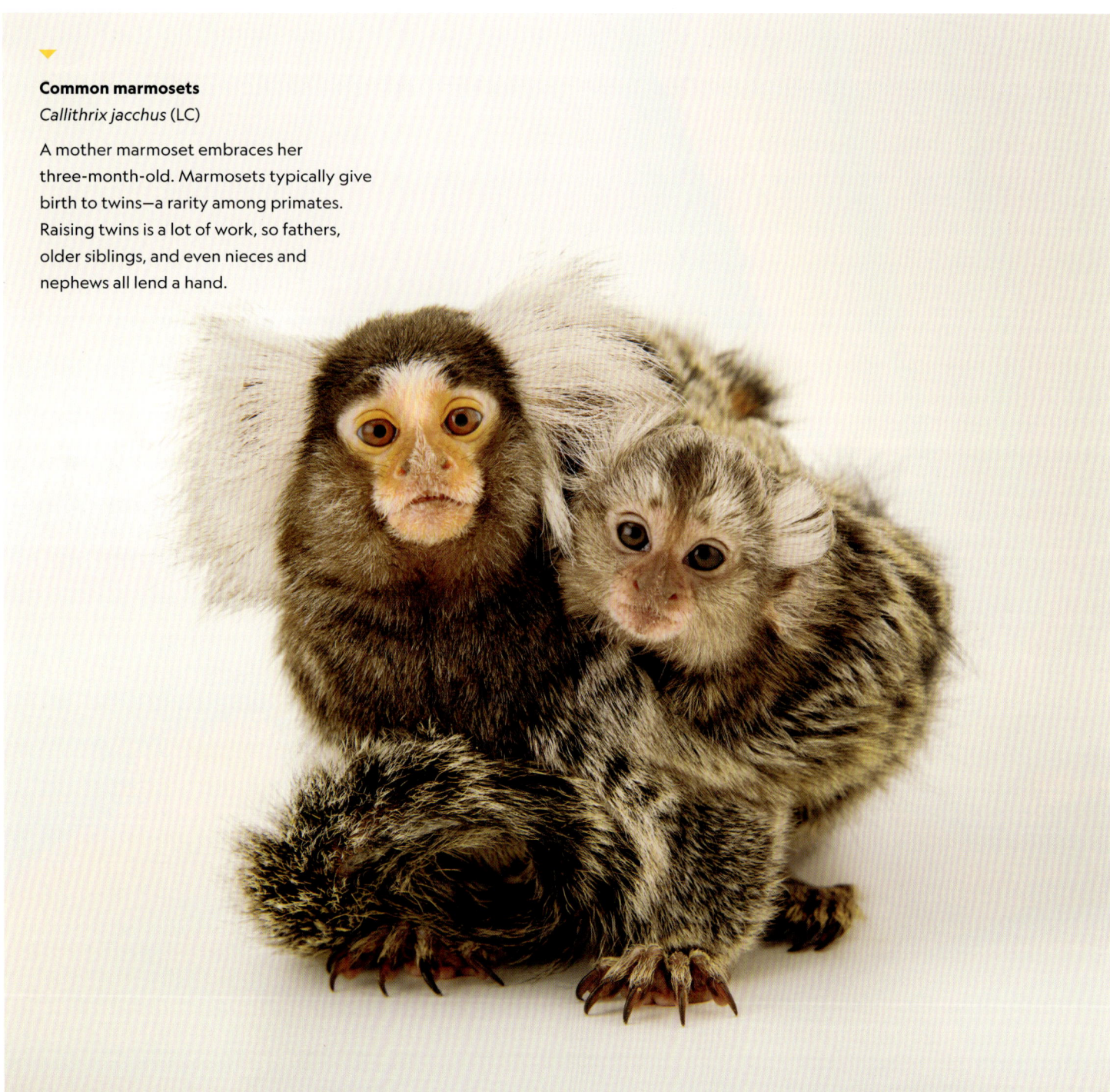

Indochinese silvered langurs

Trachypithecus germaini (EN)

These two young'uns belong to a leaf-eating species native to Southeast Asia. Infant langurs have bright orange fur, which helps their moms see them in the forest. The orange fades to silver as they grow.

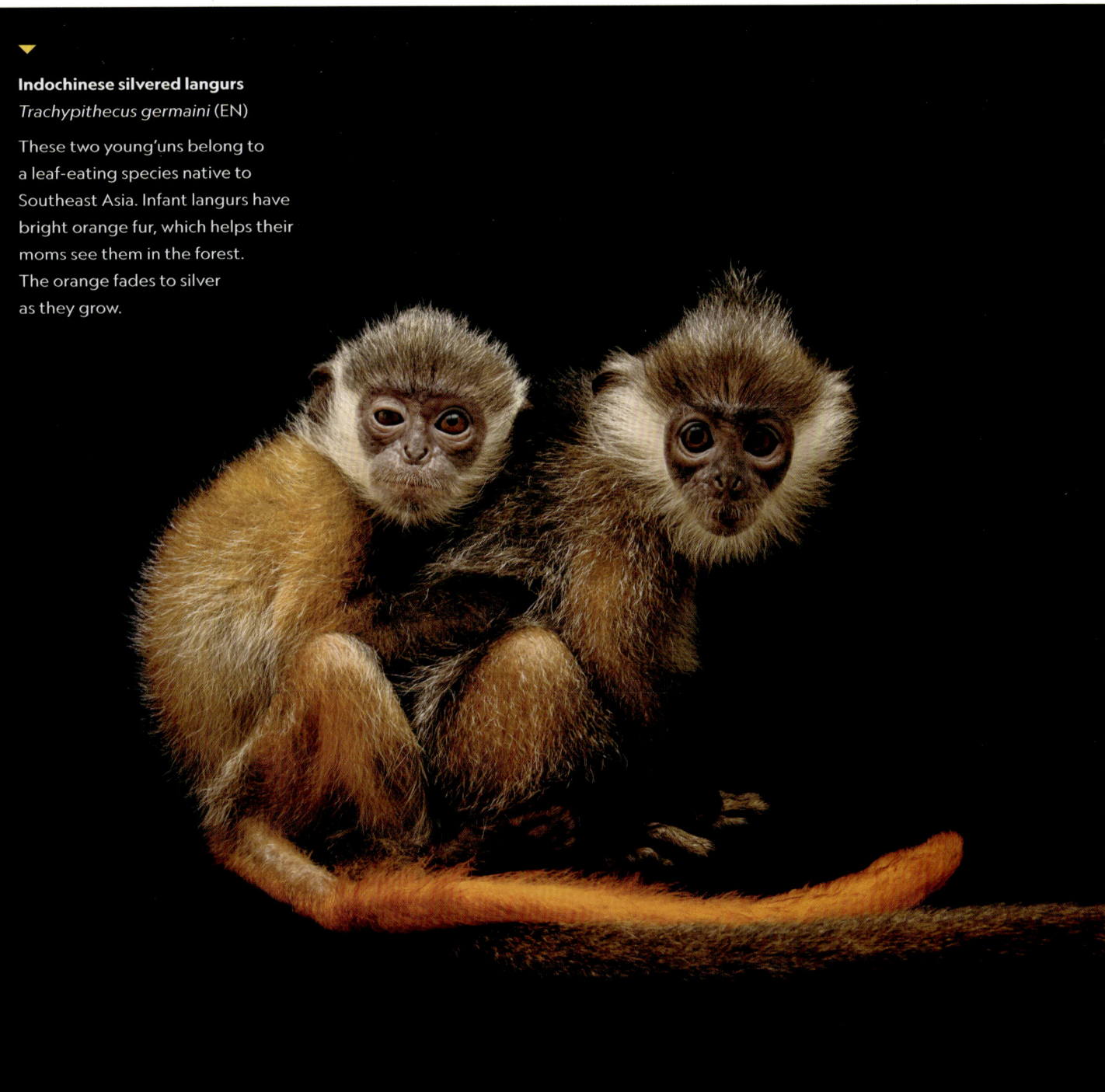

Pygmy hippopotamuses

Choeropsis liberiensis liberiensis (EN)

Females give birth on land to a single calf and leave it parked in an out-of-the-way pool of water, returning periodically to nurse the baby. After three to five months, the young start moving around and following their mom.

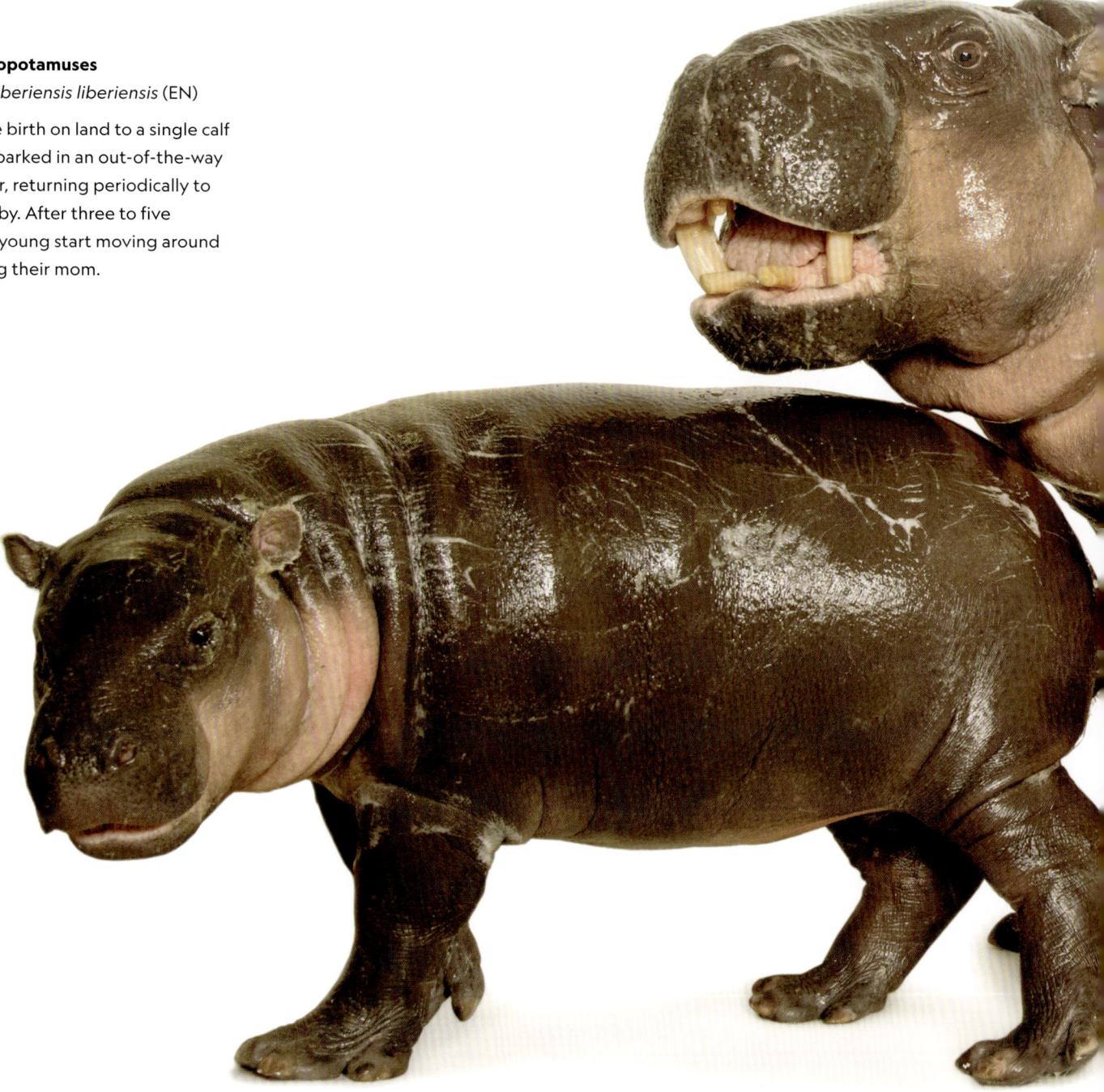

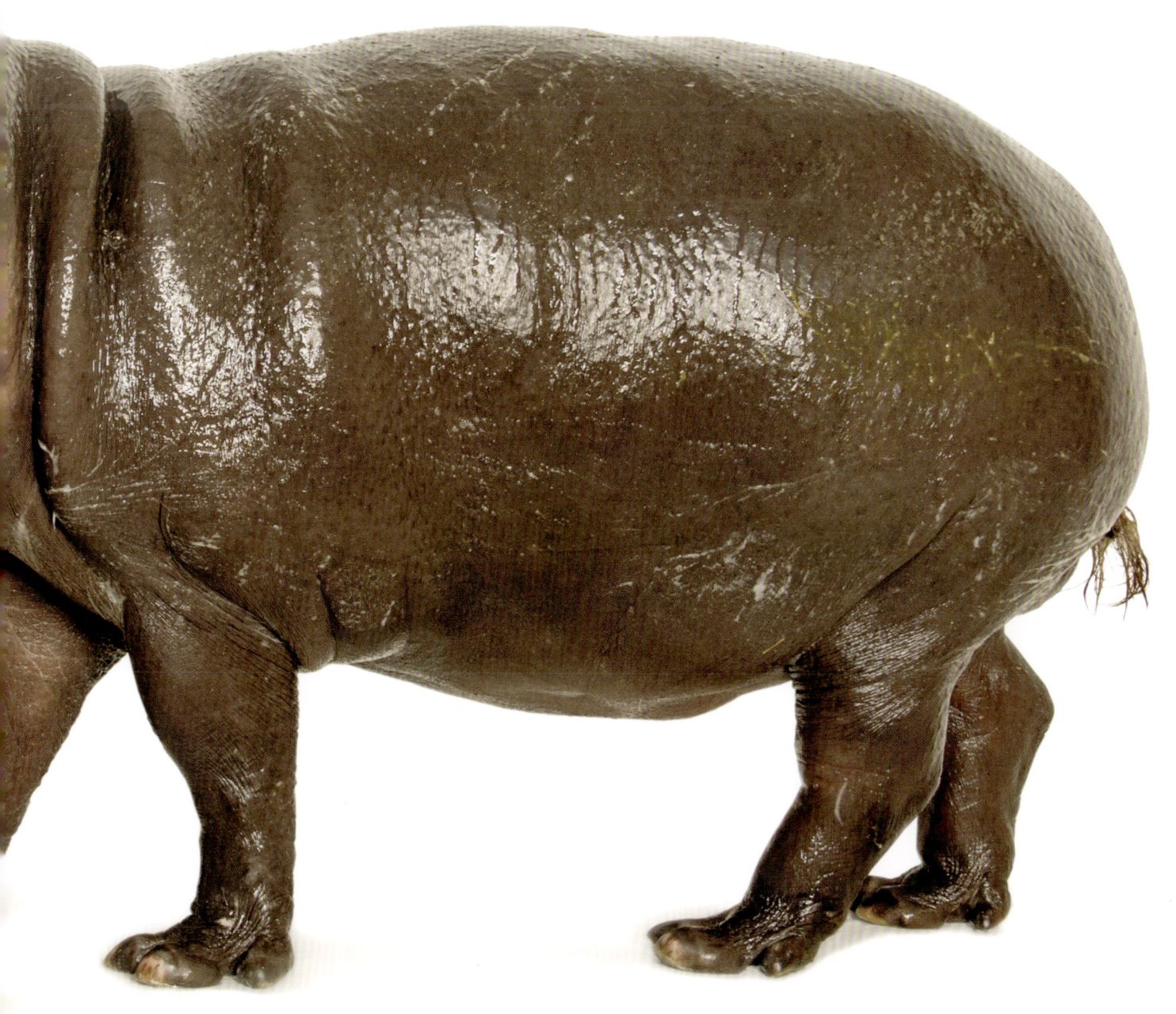

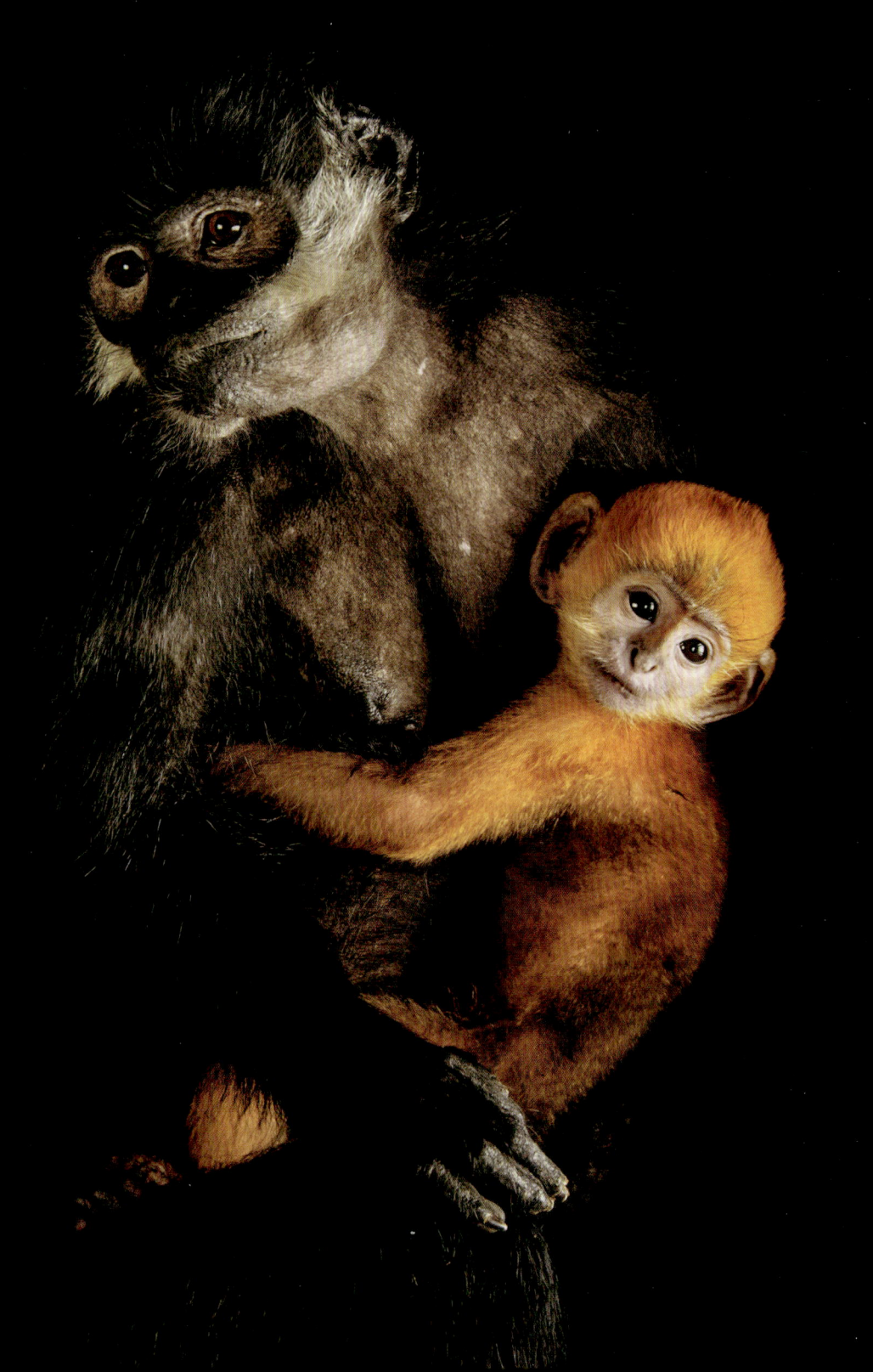

François's langurs
Trachypithecus francoisi (EN)

As langurs mature, their fur darkens. Within a year, the brilliant color of young langurs will have turned black—except for a distinctive strip of white along their cheeks.

BRIGHT SPOT
Keeping an Eye Out

T

THE AGILE, acrobatic François's langur practices alloparenting, a behavior in which adult female langurs care for the babies of the group while the mothers venture out to find food. When a jet-black mother forages along the rocky karst formations of northern Vietnam and southern China, she needs some way to spot her baby among the shadowy gray cliffs, even though the baby is in the care of other adult female monkeys. A flash of bright orange fur on her little one is just what she's looking for. The babies' striking color makes them more visible to their parents and other adults when it comes time for the young monkeys to learn to navigate the treacherous cliffs.

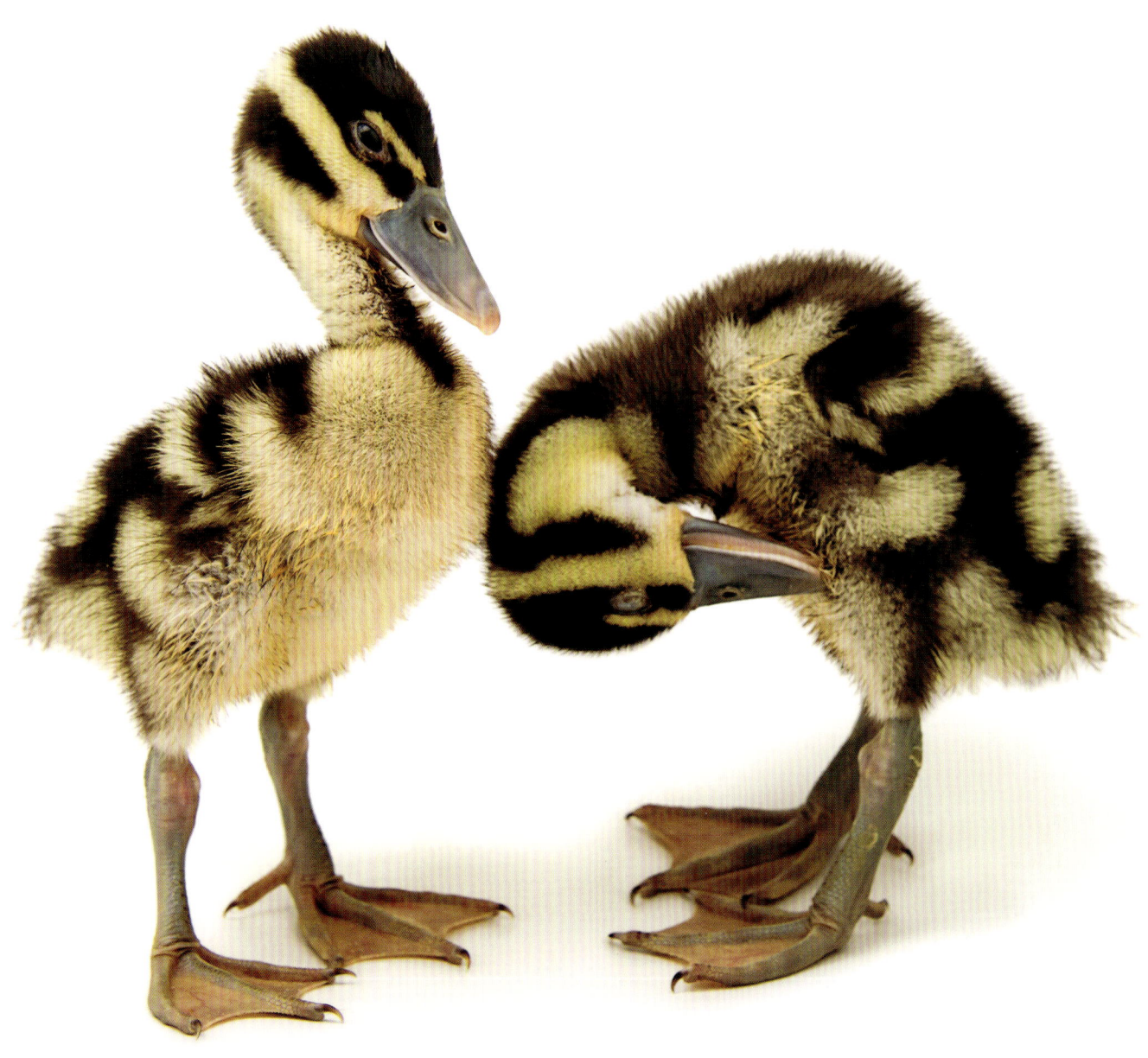

Black-bellied whistling-ducklings
Dendrocygna autumnalis (LC)

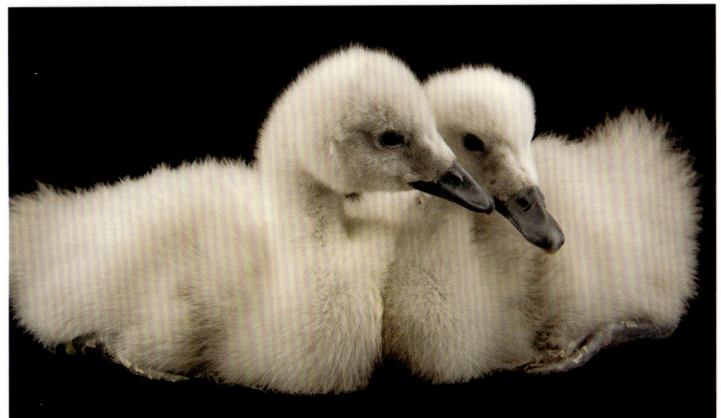

Desert cottontail rabbits *Sylvilagus audubonii vallicola* (LC)

SIBLINGS FOREVER

The only thing cuter than one baby animal is two. Throw in some cuddles and a red panda kiss, and we're goners. The science on cuteness is clear: Humans are hardwired to respond to big eyes and round bodies because they remind us, on a primal level, of our own offspring. Add in some soft fur and fluffy down, and we can't resist.

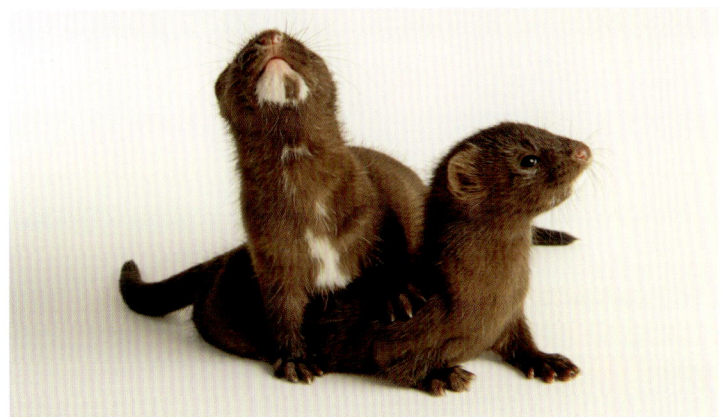

Black-necked swans *Cygnus melancoryphus* (LC)

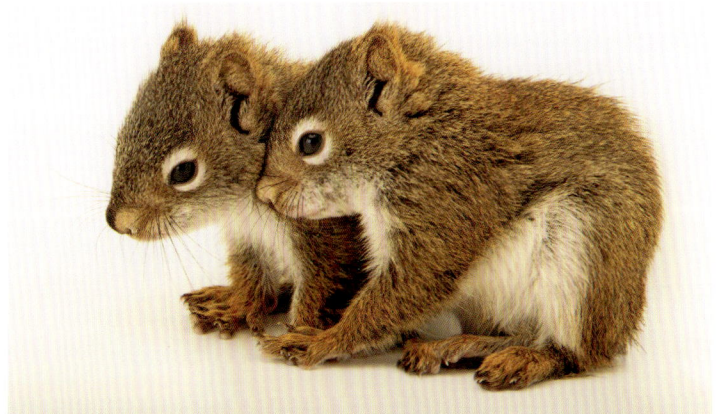

Fremont's American red squirrels *Tamiasciurus hudsonicus fremonti* (LC)

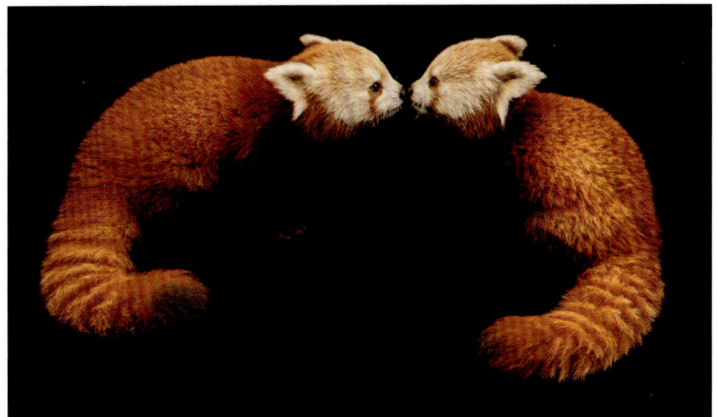

American minks *Neovison vison* (LC)

Western red pandas *Ailurus fulgens fulgens* (EN)

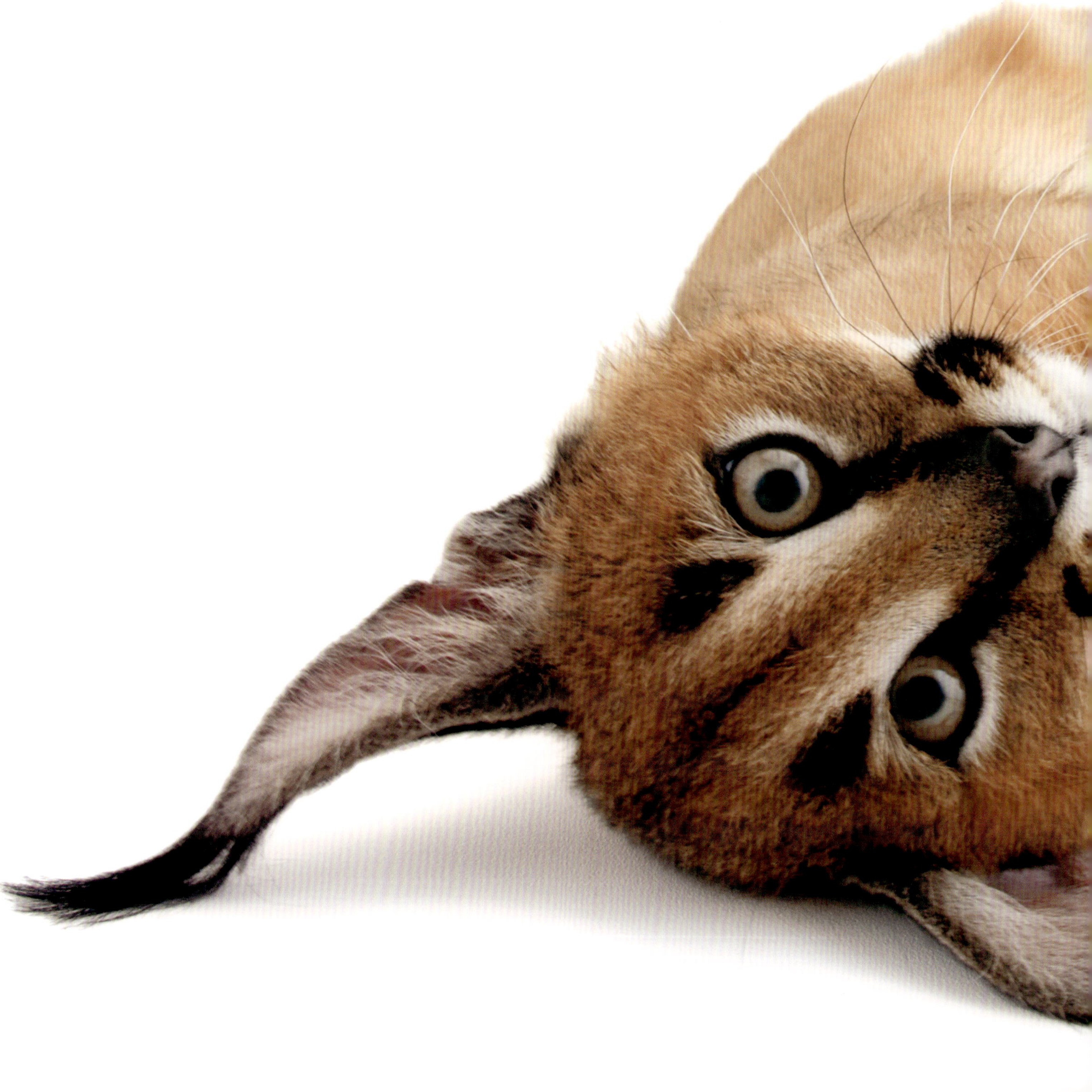

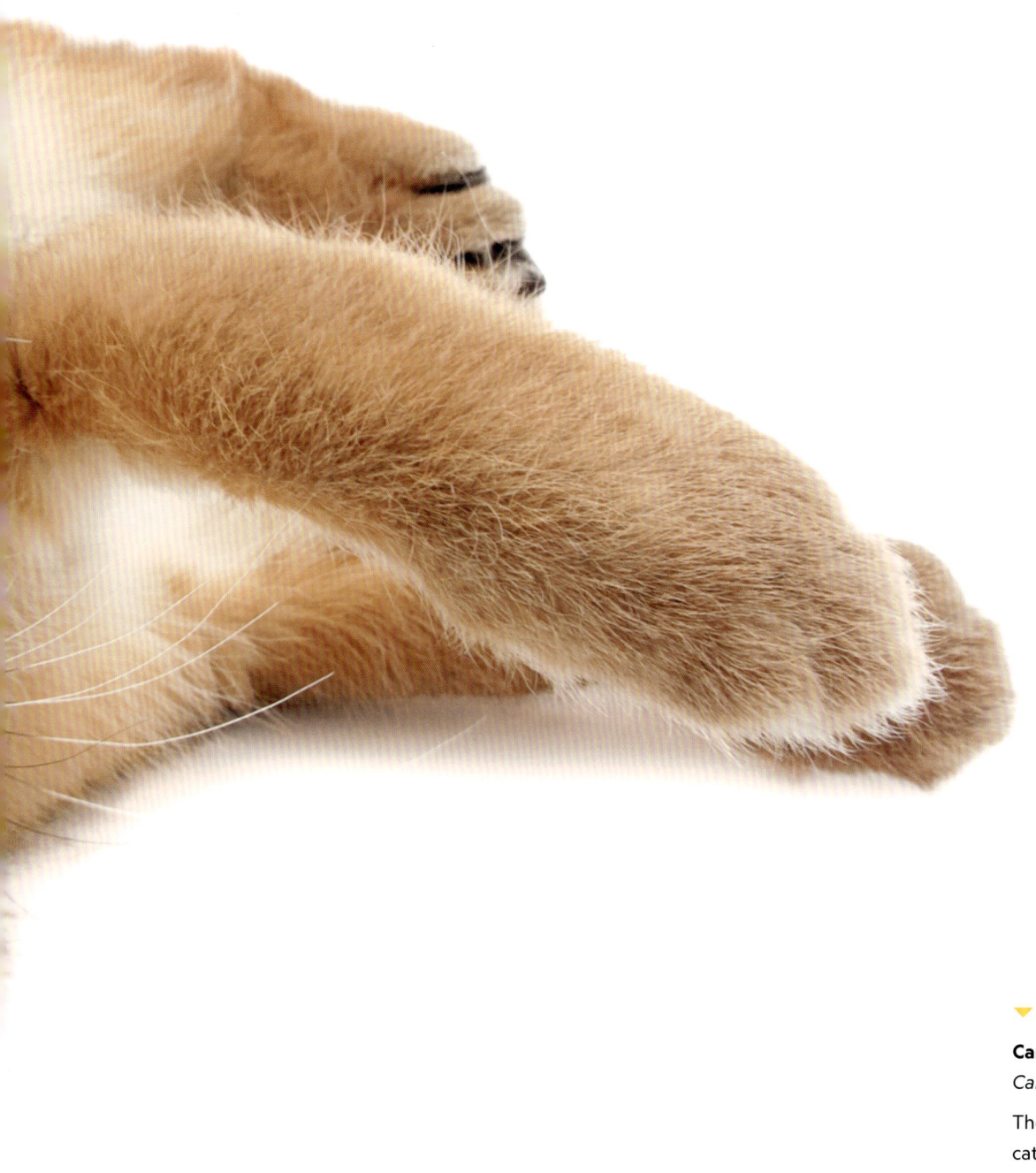

Caracal

Caracal caracal (LC)

The largest of the small African cats, adult caracals live mostly solitary lives. But kittens, like this three-month-old, are playful and snuggly and will stay with Mom until around a year old.

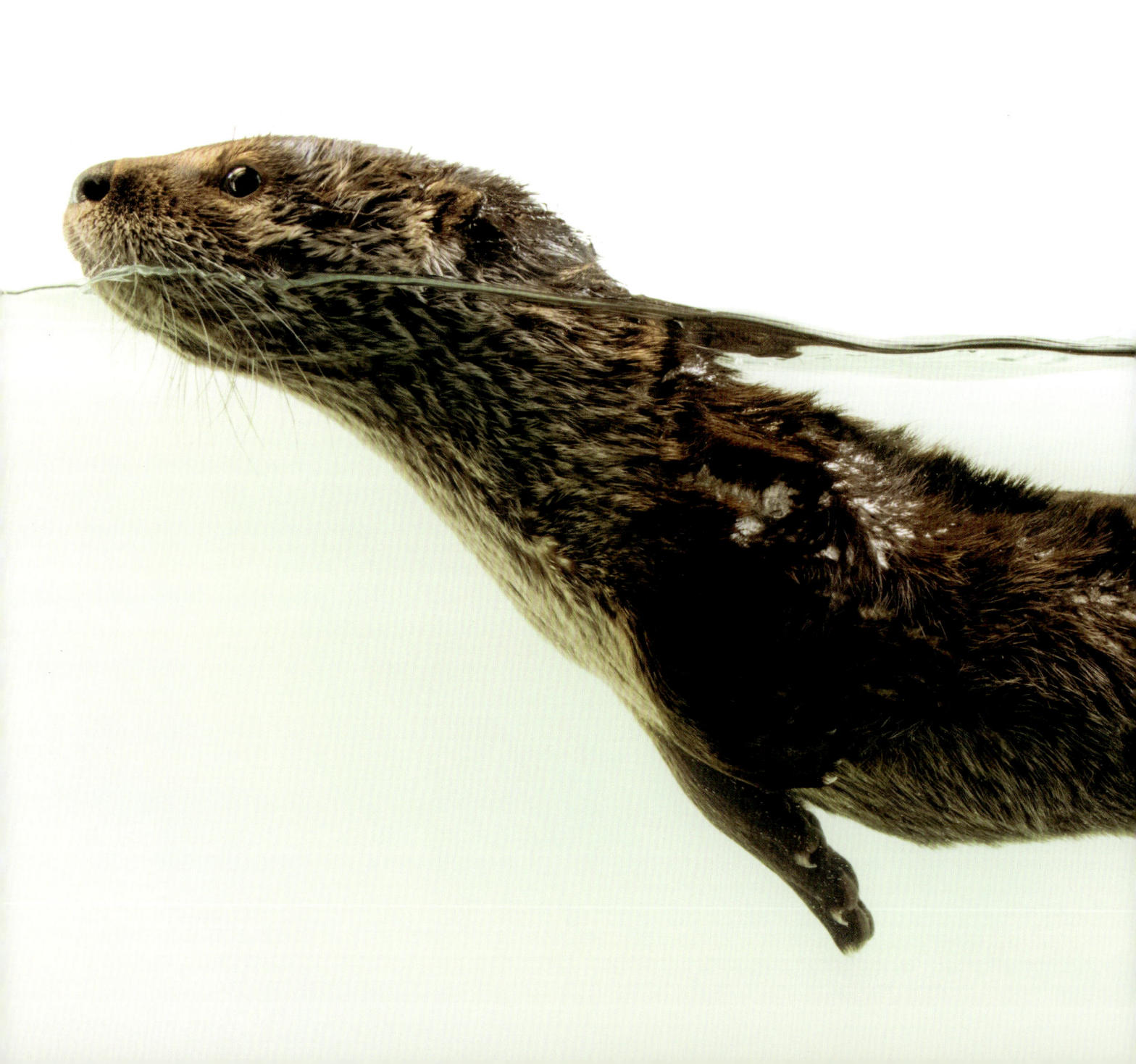

5 | FLEDGING

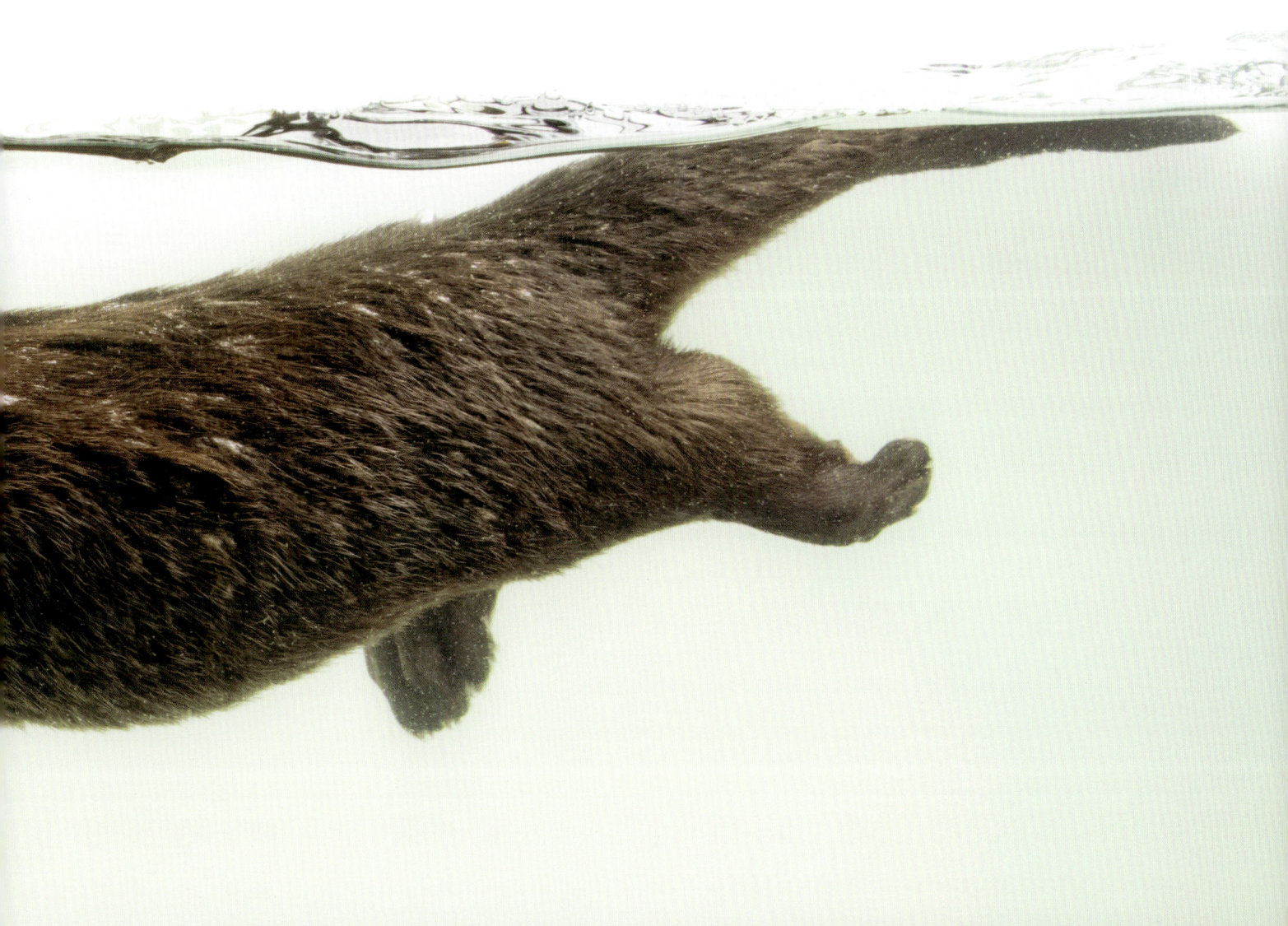

You're on Your Own, Kid

As animals draw closer to adulthood, their instincts become powerfully honed with know-how that is a combination of innate abilities, sharp senses, and learned behaviors. Their exploits often make us wonder: How do these youngsters know what to do?

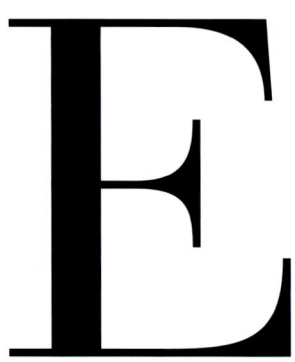

EVERY YEAR AS SUMMER WANES, young birds just a few weeks old prepare for an arduous journey. Many fatten up and fly thousands of miles across oceans and mountains and cities, bound for places they've never seen. A few will become the ultramarathoners of the bird world: Soon after they hatch, Arctic terns fly from pole to pole, an annual round-trip distance of more than 25,000 miles (40,000 km). Nobody gives migratory birds directions or shows them where to go; often they fly alone. Scientists have discovered that the stars, the sun, and Earth's magnetic field all help young birds carry out this incredible triumph of navigation, but the full story of what guides them is still a mystery.

Migration is one impressive feat, but there are others: Seabirds like the Hawaiian petrel fledge from their cliff burrows above the sea when they are only a few months old and have shed their fluffy gray down for sleek oceangoing feathers. Straight out over the open ocean they go; they will not touch land for several years. Only when they are ready to breed do they return—often to the very

same place where they hatched. How do they know where to go, or when to come back? Another mystery.

Birds aren't the only ones so brave. When a green turtle cracks open its egg, it crawls free and immediately dashes for the ocean, guided only by instinct and light from the stars and moon. The hatchling is all business as it flips and flops its way to the surf; predators hover, ready to swoop in and pluck a snack from the wave of tiny turtles scrambling for the sea. If a baby turtle reaches the water—most do not—it has only traded one set of perils for another. Now it must make its way far out into the open ocean, where it will live for a time before embarking on yet another long journey to its coastal foraging grounds.

Daring travels are one way young animals show their fortitude. Grit is another: Some mature quickly, hardwired to master key behaviors while only a few weeks or months old. Snow leopard cubs are born blind and helpless in the rugged mountains of Central Asia. They begin following their mother when they're only three months old. For just under two years they stay with her and play with their siblings, learning everything there is to know about life in the Himalaya. The family bond is strong but temporary. Adult snow leopards are solitary, and eventually these ghosts of the highest mountains on Earth part ways. On silent, padded feet, they disappear, alone. They may never meet again.

Eastern grass owls *Tyto longimembris chinensis* (LC)
Previous pages: **North American river otter** *Lontra canadensis* (LC)

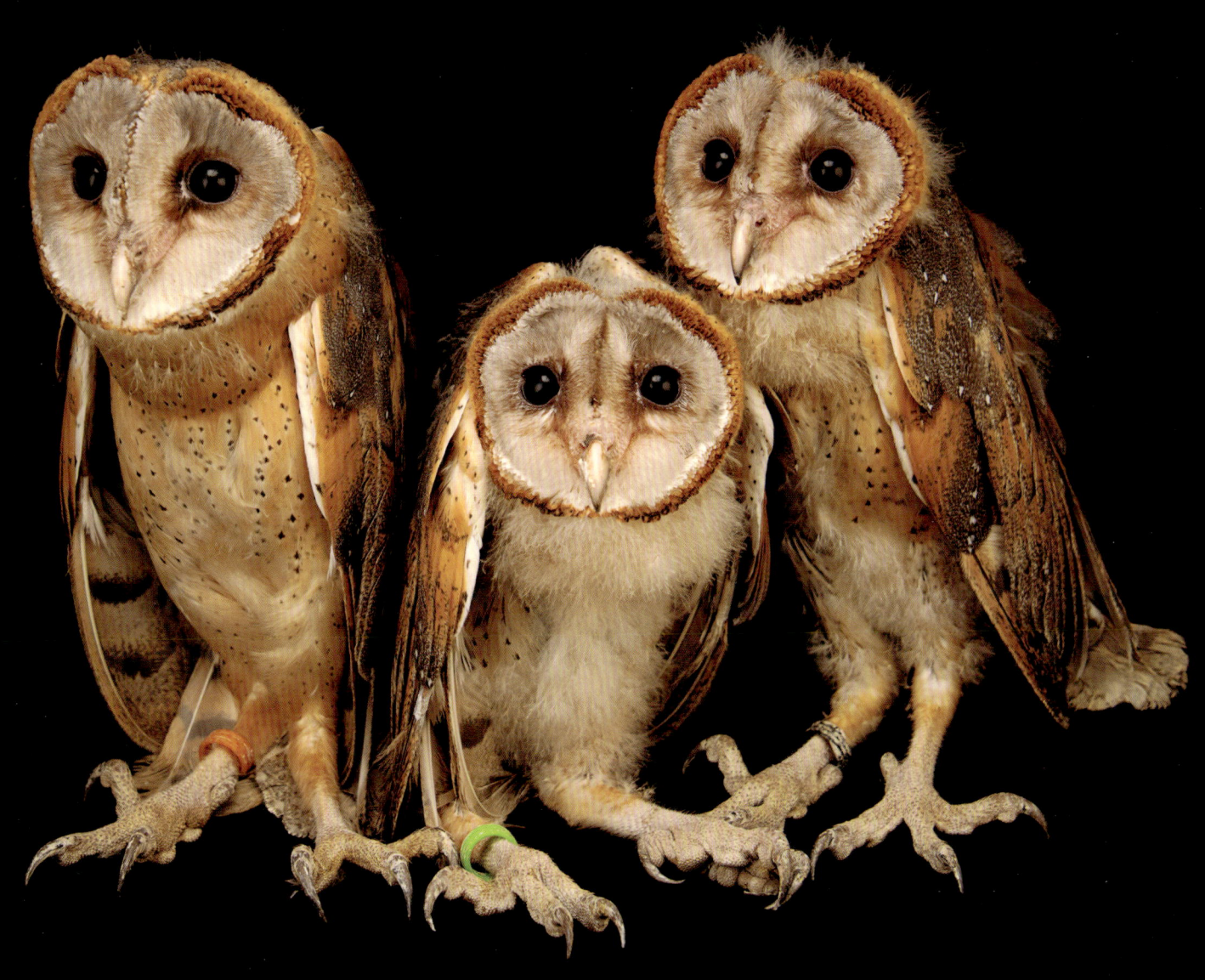

Tiny Tigers
Last of Their Kind

T

THE CRITICALLY endangered Scottish wildcat is one of the rarest animals in the United Kingdom and the only surviving native wildcat. Although it looks like a domestic tabby cat, the wildcat, also known as the Highland tiger, is much stockier and more muscular.

Wildcat kittens are born in a litter with up to seven siblings. They nurse from their mother for about a month, and she cares for them for another five months until they are fully independent. The kittens are blind when they're born, but their lack of sight doesn't mean they aren't quick learners. Their mother brings them live prey, and by the time they're nine months old, these precocious young felines, full of sass, are excellent hunters.

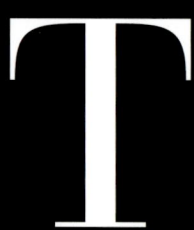

Scottish wildcat
Felis silvestris grampia (CR)

This wildcat once roamed the woodlands of Britain. Born helpless, the kittens quickly become skilled hunters.

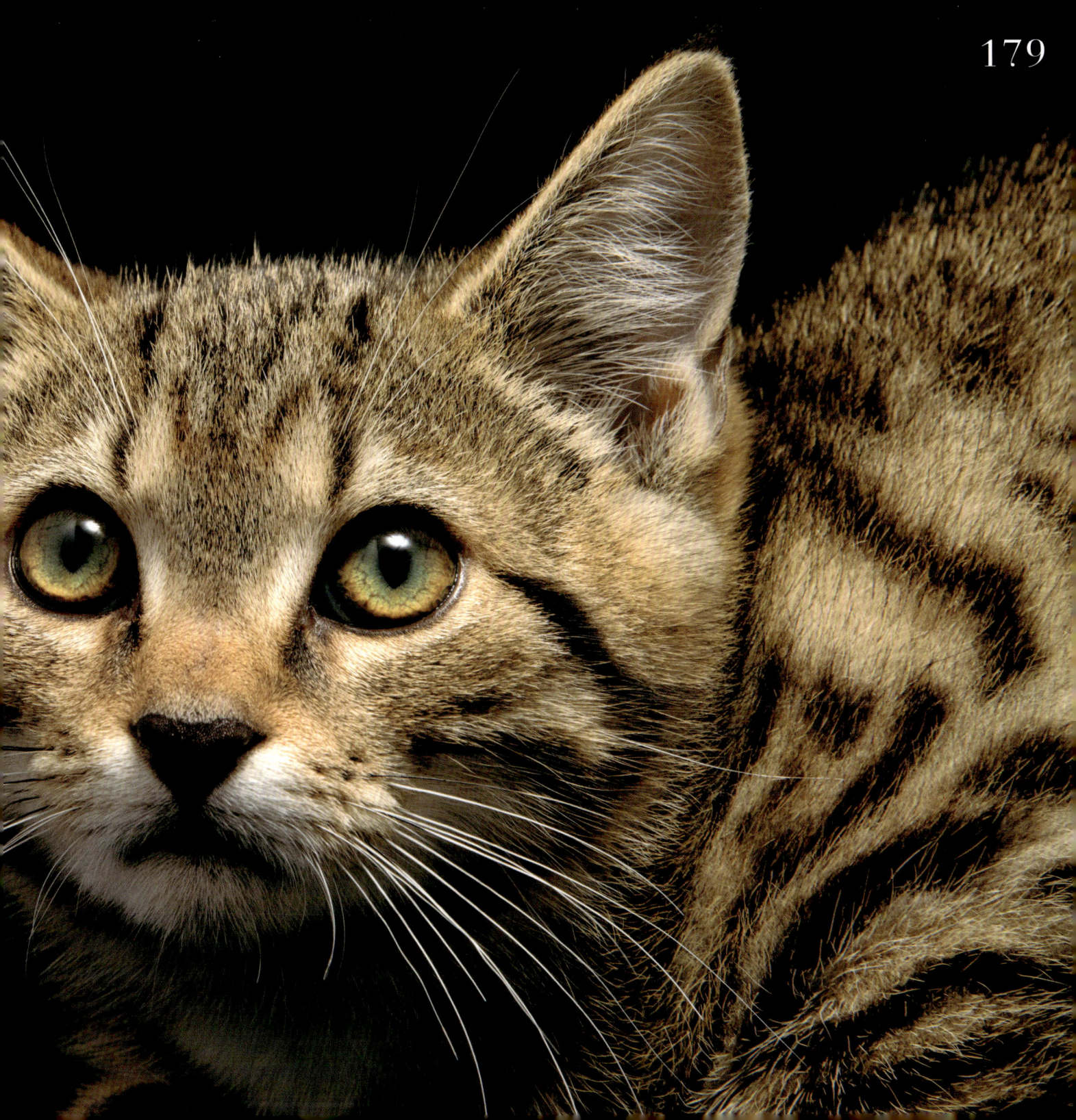

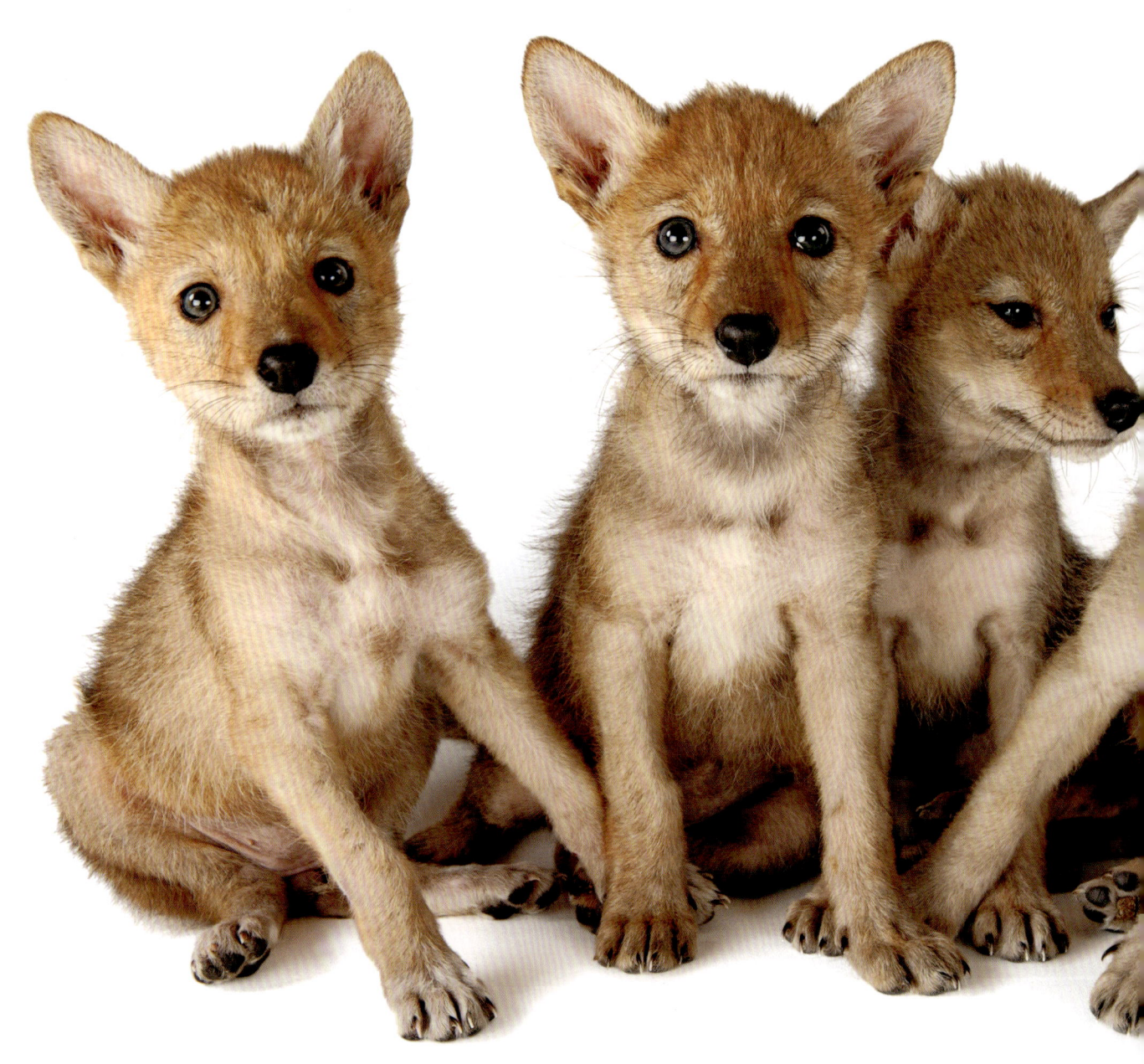

Coyotes

Canis latrans (LC)

Coyotes are born in the spring in a den. Doting parents feed and care for the pups. After six weeks, the pups begin exploring, and they're hunting alone by summer's end.

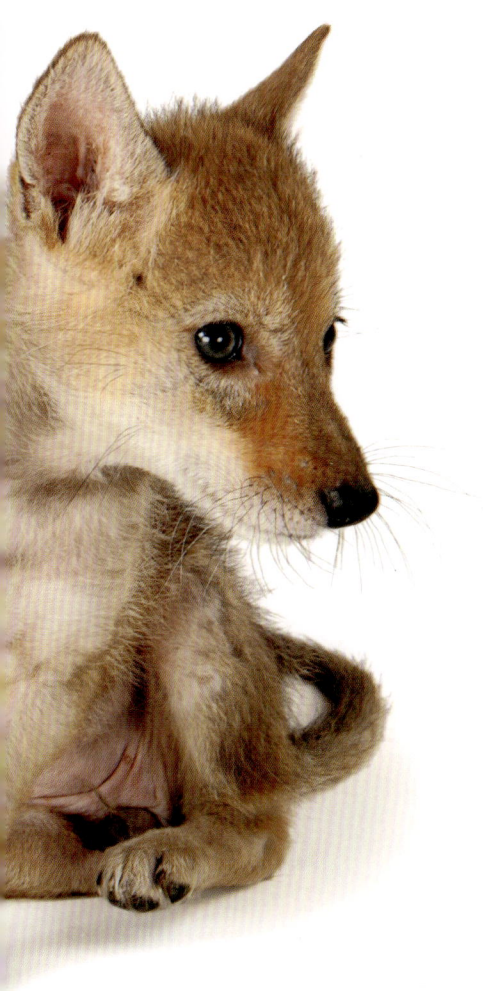

BEHIND THE LENS

EVERYBODY FREEZE

ANIMAL: Coyote, *Canis latrans* (LC)

LOCATION: Lincoln, Nebraska

"Friends of mine who work as wildlife rehabbers brought four coyote pups to a photo studio I used to have in the spare bedroom of my house. I decided it would be cute to do all four coyote pups at once. But the only way to do it would be for my friends to take all four and line them up abreast. We'd count to three, they'd let go, and I'd snap the picture.

"The pups just sat there for a minute—and then they shot out, scampering all over the second floor. They couldn't navigate stairs very well, so they couldn't go up to the third floor or down to the first floor, but they ran into all the bedrooms. So on the second try, we closed the doors to the bedrooms. They went into the bathrooms instead. It was mayhem. We did that two or three times—it might have been more—till we got one photo that looked perfect. They all looked cute; they were all attentive.

"You can distort reality a bit with photography because it freezes time so well. In that moment, they were all sitting there looking calm and patient, like willing subjects. But they were ready to run—not because they were scared, but because they were pups excited to be exploring new territory."

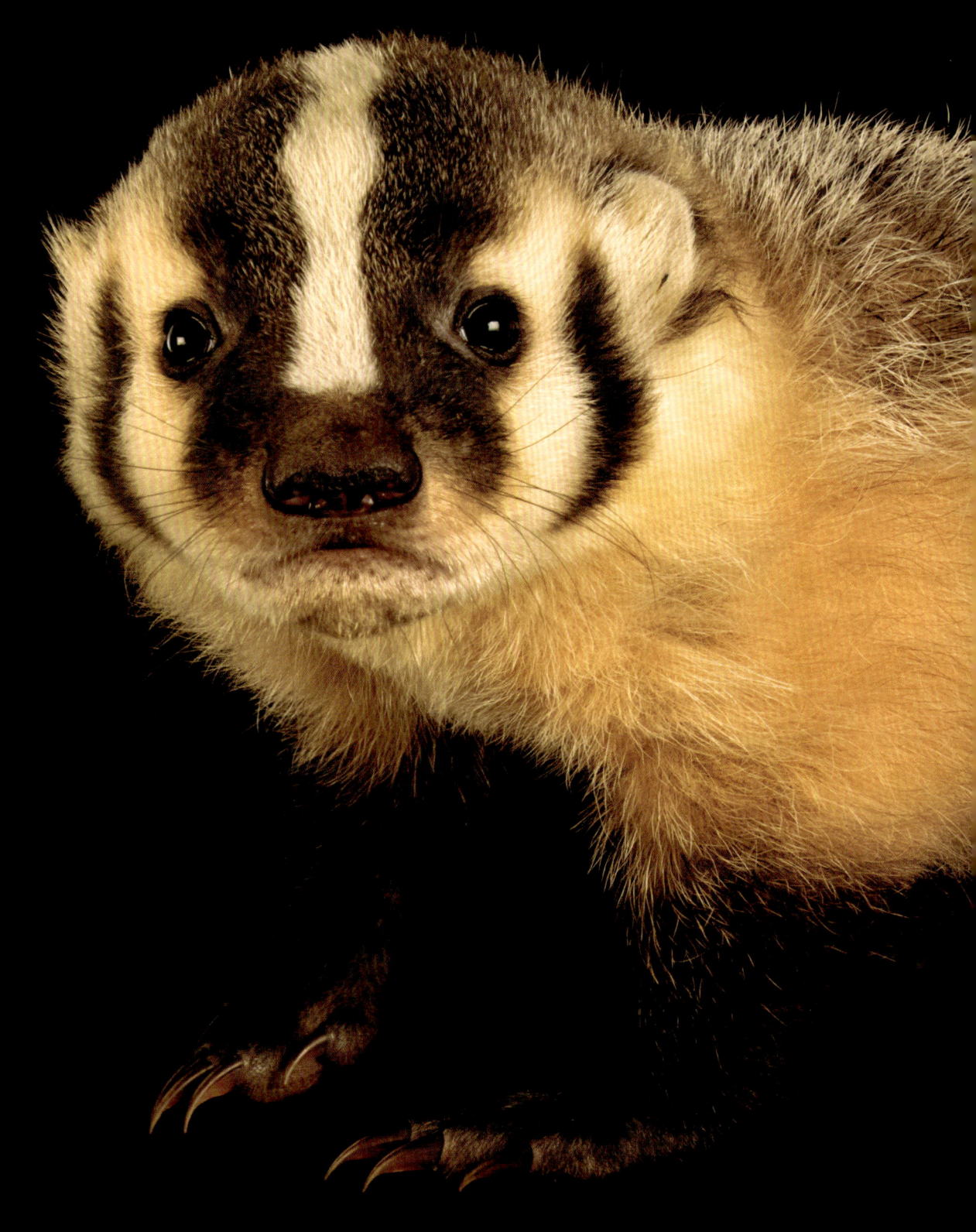

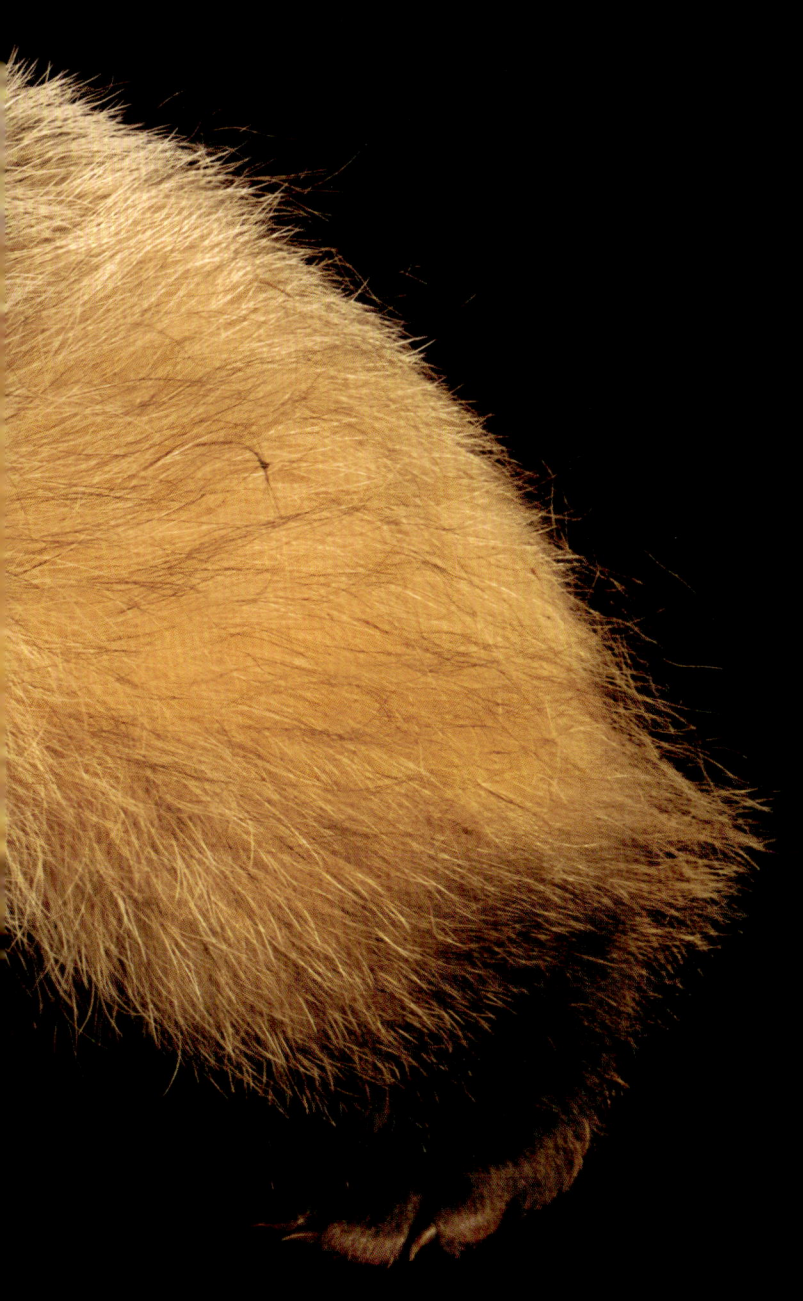

▼

American badger

Taxidea taxus taxus (LC)

Badger mothers give birth in the spring to a litter of several babies. The little ones grow quickly: They emerge from the den after about six weeks and go out on their own by six months of age.

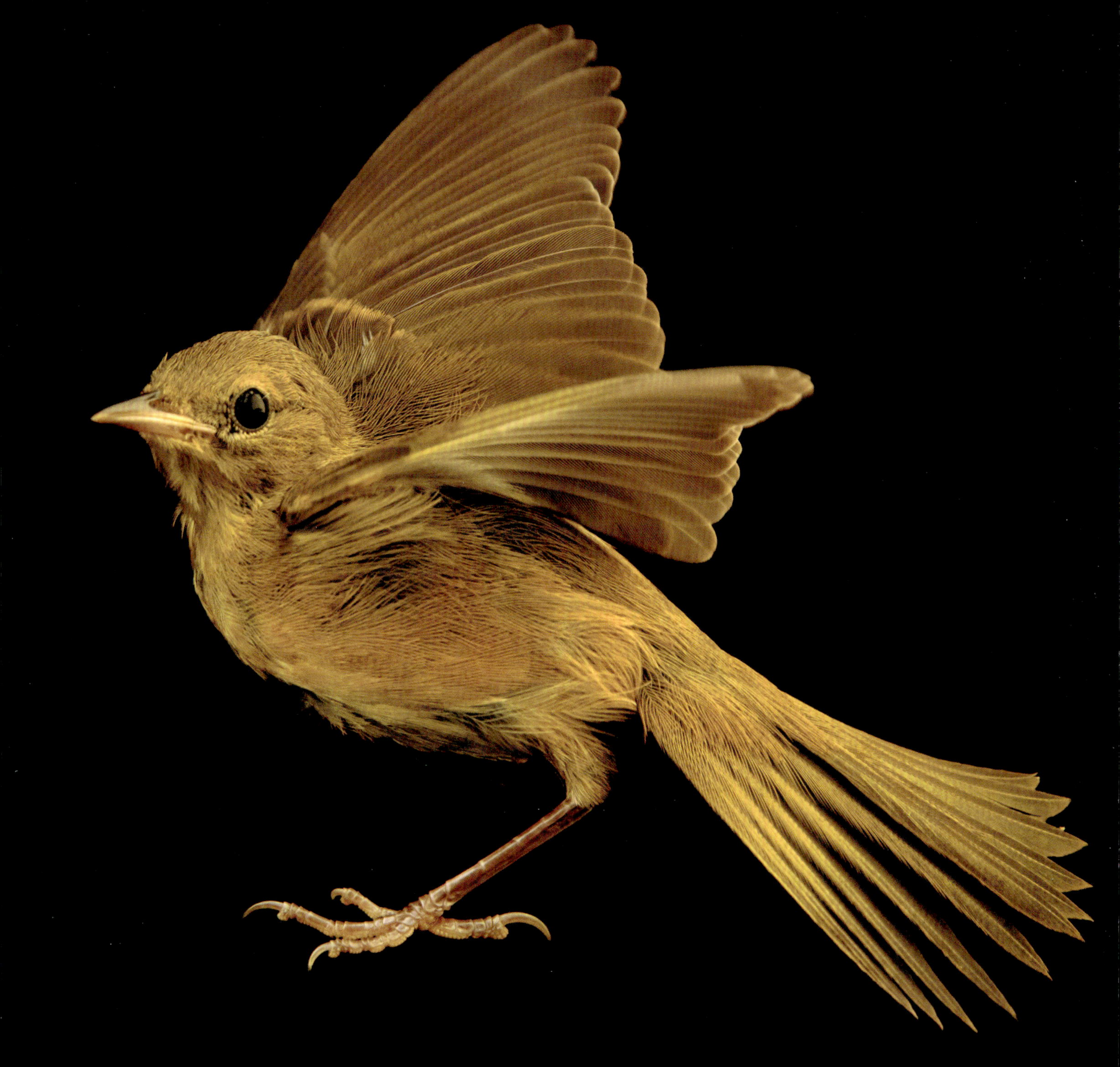

> 66
>
> "I have seen remarkable recovery in some species that were nearly extinct. And now they are out of the woods, all because of people who cared."

▼

Common yellowthroat
Geothlypis trichas (LC)

Within the first week of this little warbler's life, its mass increases fivefold, and it sprouts feathers and opens its eyes. By day seven, it is nearly ready to leave the nest.

SAFE PASSAGE
First Flight

H

HELPLESS AND FEATHERLESS is how this small warbler greets the world when it hatches in early summer. Its skin is dark orange with only a few tufts of grayish down. It cannot even lift its head. There is little hint that in just a few weeks this floppy, naked nestling will embark on a perilous journey from northern Canada to Central America.

Within two weeks of hatching, the fast-growing warbler chick will leave the nest, climbing out and taking cover in nearby thickets. Over the next few weeks, the youngster eats well and practices flying. By late summer, the young bird is ready. It feels the seasonal restlessness common to migratory birds as their bodies yearn to travel. Finally, the little warbler takes to the skies and, along with so many other species, turns south, one of billions.

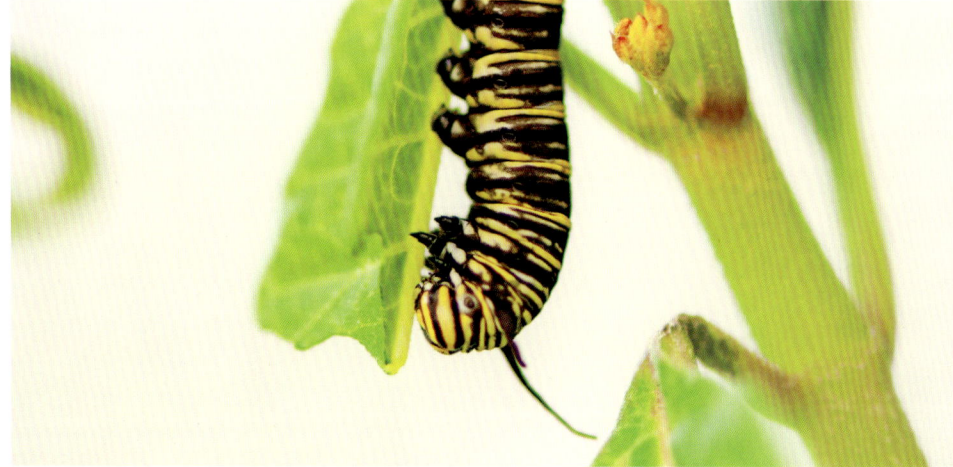

Monarch butterflies
Danaus plexippus (LC)

After nine to 15 days of completing its metamorphosis inside the chrysalis, an adult monarch butterfly emerges. It now begins a short life focused on finding food, migrating, mating, and, if female, laying eggs.

SAVING THE SPECIES

EPIC JOURNEYS

One of the most intrepid journeys among any newly minted animal is accomplished by an insect. Most North American monarch butterflies spend the winter in high-altitude fir forests in Mexico and pass the summer in the northern United States and southern Canada. As scientists study the butterflies' migratory patterns, they gain a clearer picture of what kind of habitat the insects need at different stages of their life cycle.

Butterflies that overwinter in Mexico head north in the spring and lay eggs on native milkweed plants along the way. The eggs hatch, the caterpillars soon zip themselves into their chrysalides, and before long comes the next generation. The entire life cycle takes about four weeks and continues as each subsequent generation of monarchs works its way northward. The last monarchs to hatch along the migration route break free of their chrysalides in southern Canada in late summer. These monarchs are different from their predecessors. They will live eight times longer and travel much farther: Scientists call them the "super generation." When the weather starts to cool, they turn south, bound for Mexico and their ancestral fir forests. Traveling at the speed of a bicycle, they complete in one generation a 3,000-mile (4,800 km) journey that took their forebears three or four. Only after a winter's rest will they finally mate in spring, and the cycle starts anew.

Humans can help monarchs thrive by making sure they have the plants they need to complete their unique life cycle. At various points along their migratory pathways, butterflies look for native milkweeds in yards and parks, cities, and everywhere in between to lay their eggs. Every piece of the puzzle matters.

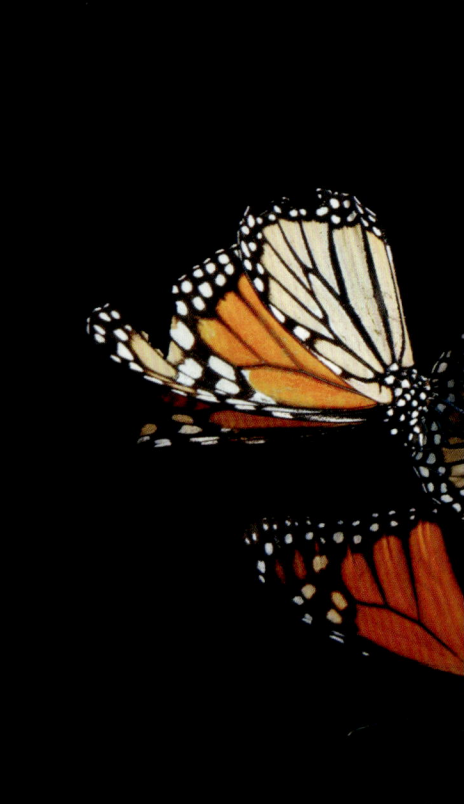

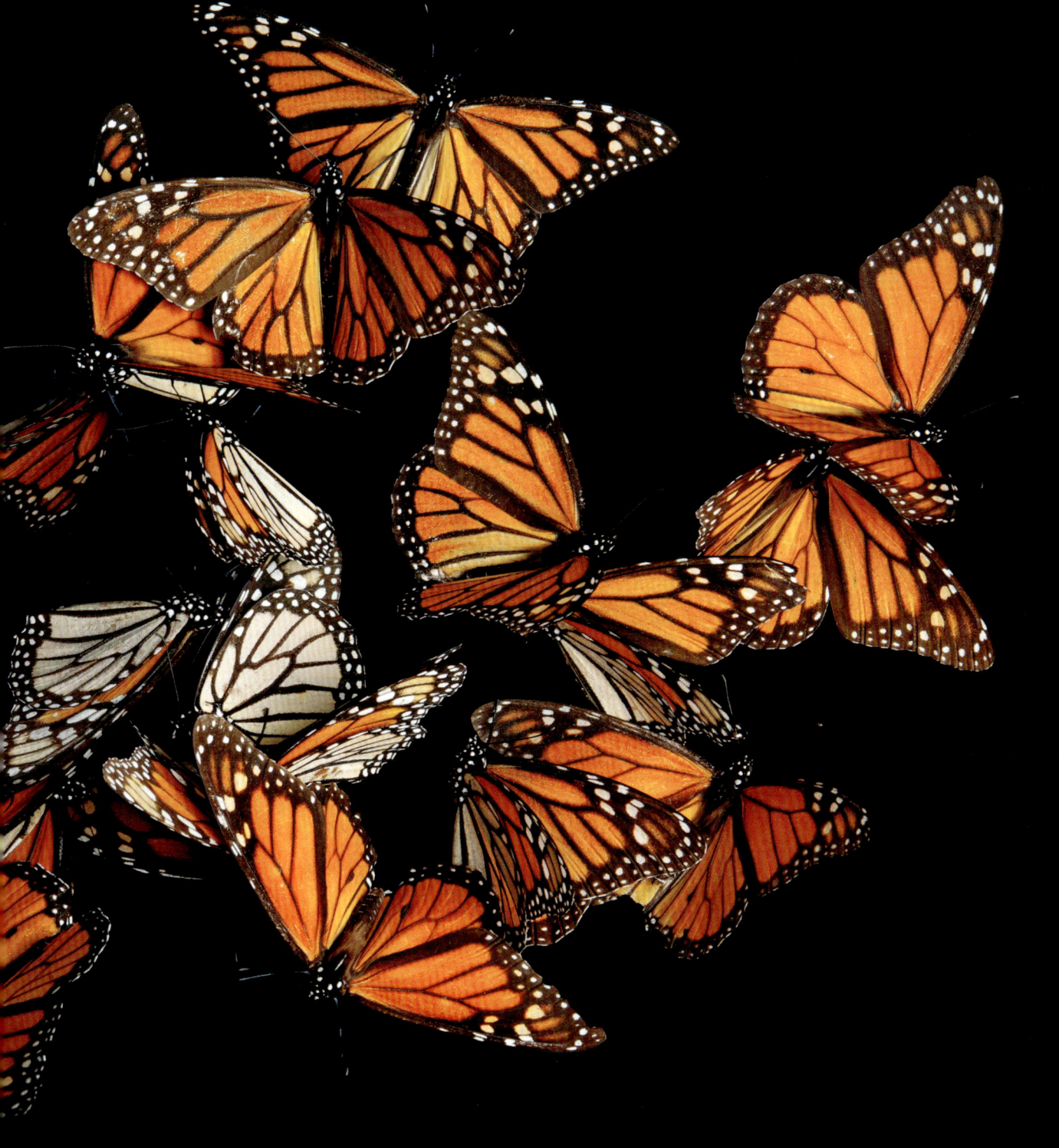

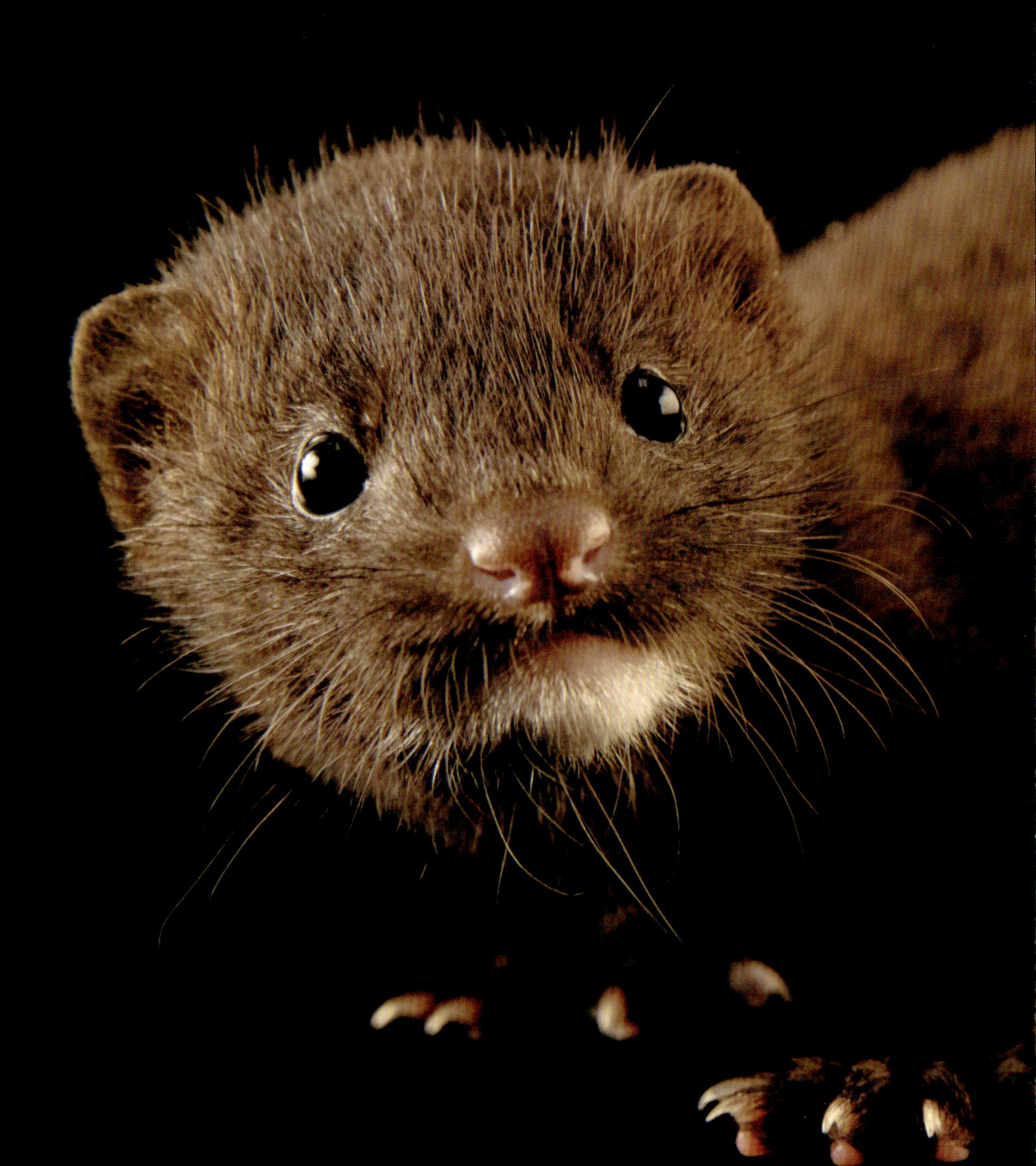

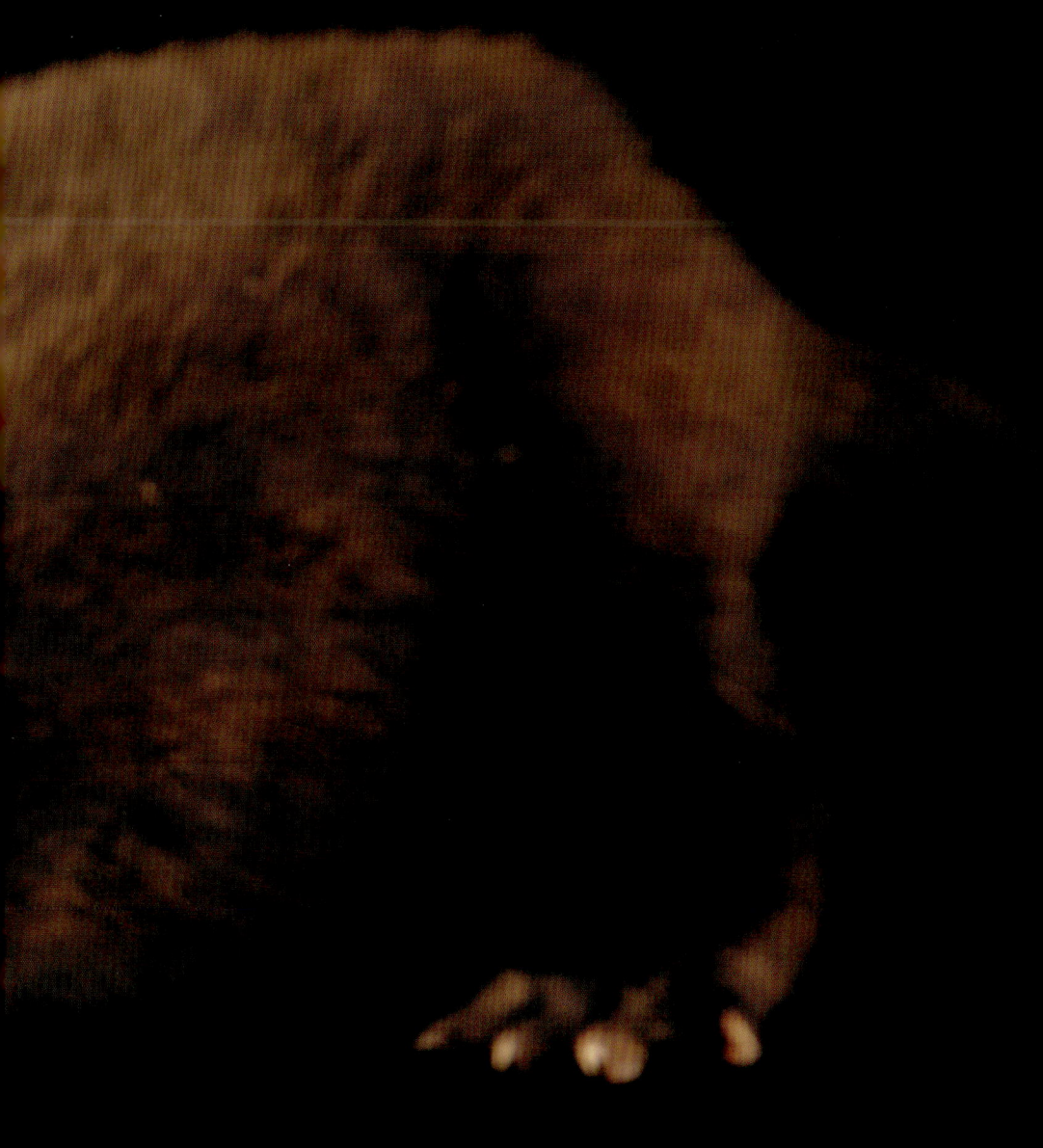

American mink

Neovison vison evagor (LC)

Spring is baby mink season.
The little ones are born tiny and
helpless, but six weeks of nursing
and a summer under their
mother's care prepare them to
go out into the world on their
own by autumn.

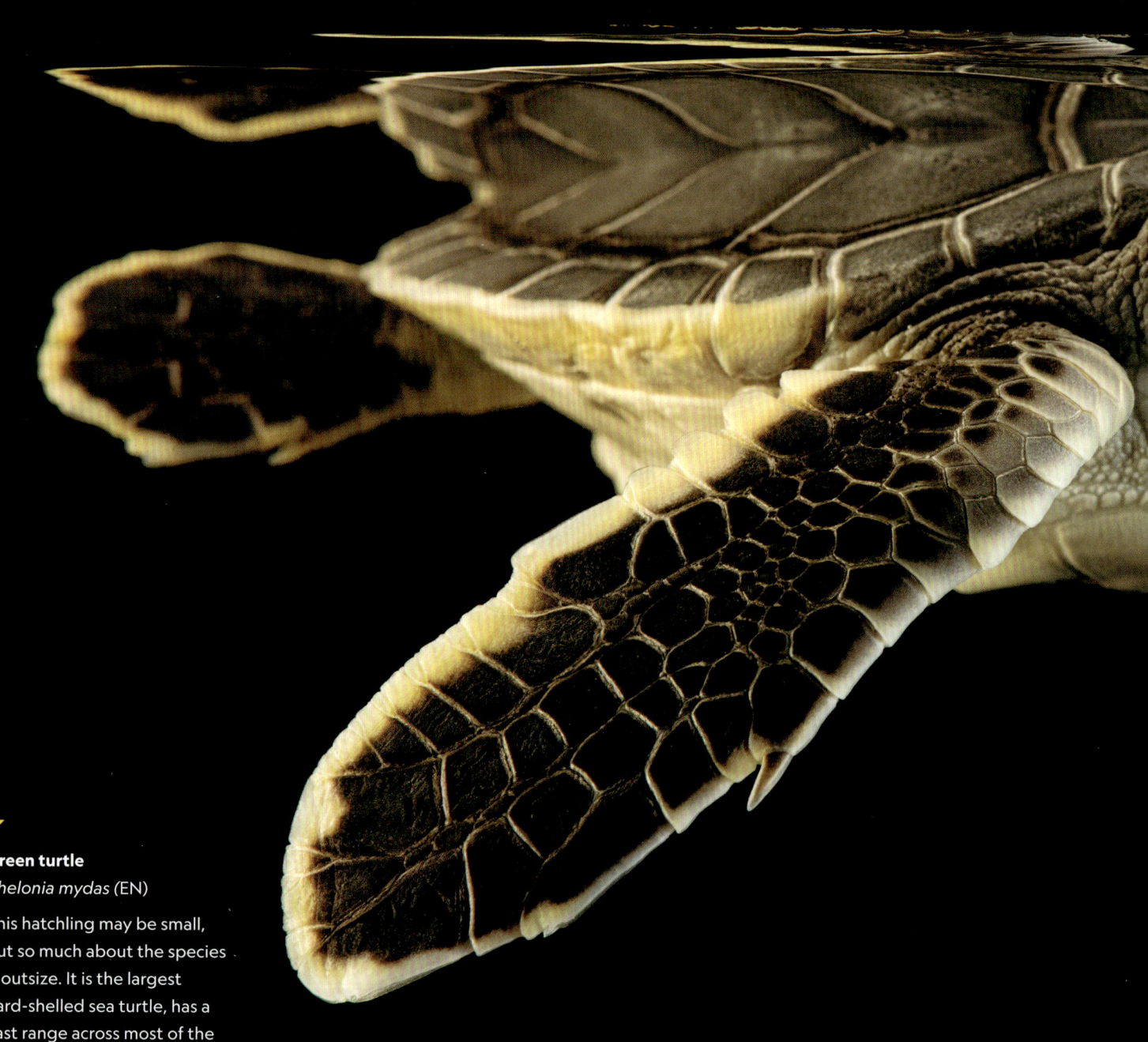

Green turtle

Chelonia mydas (EN)

This hatchling may be small,
but so much about the species
is outsize. It is the largest
hard-shelled sea turtle, has a
vast range across most of the
world's oceans, and can live up

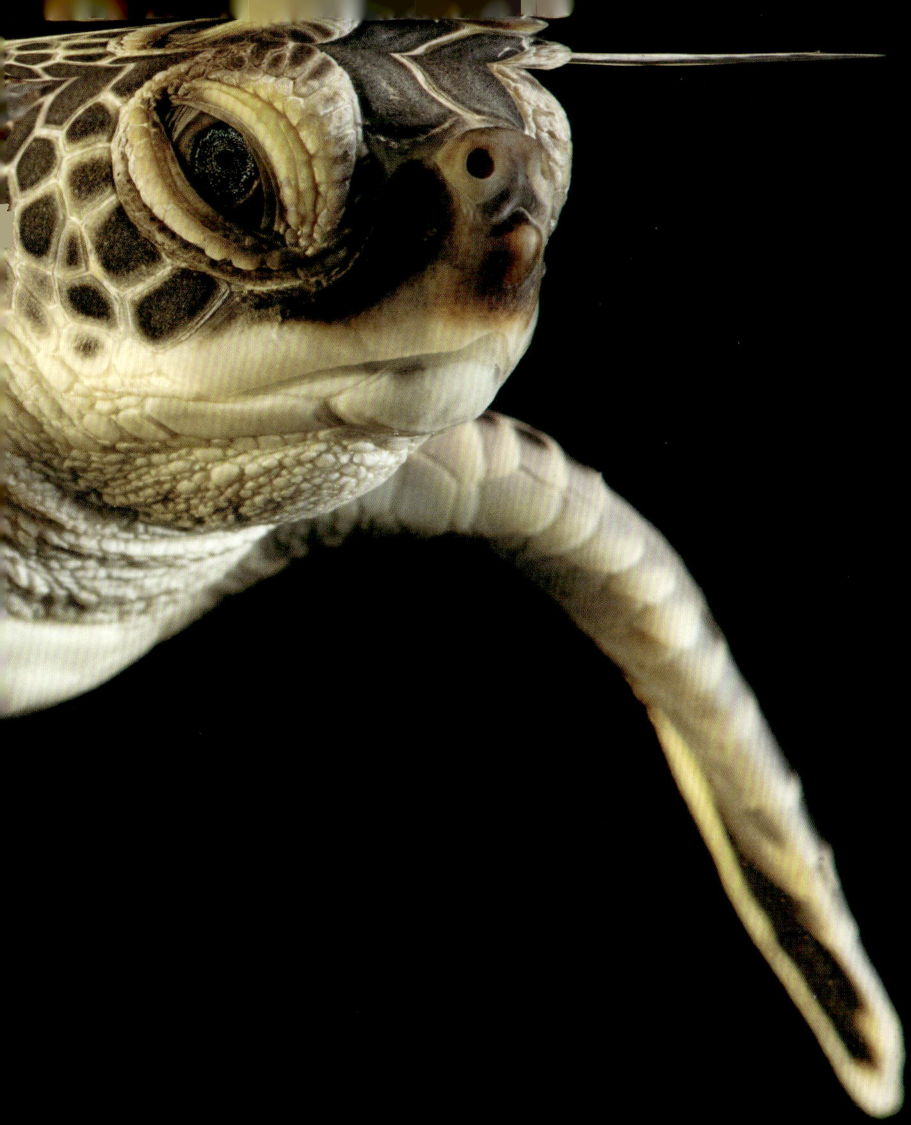

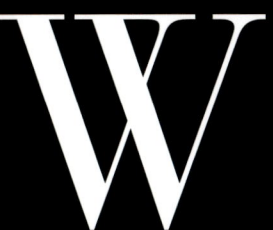

W

WHEN FEMALE TURTLES are ready to lay eggs after reaching sexual maturity, somewhere between 25 and 35 years of age, they return—sometimes traveling up to 1,600 miles (2,600 km)—to the coastal beaches where they hatched decades earlier. Using their flippers, they dig out nests in the sand and lay about 115 eggs each. Then they head back to sea.

From the moment baby turtles hatch and dash for the ocean, their existence is fraught. It's a baptism by fire for the young reptiles: They must not only steer clear of natural enemies but also avoid the trawls, nets, hooks, and traps set out by fishers. They also must navigate the hazards of busy ports and developed coastlines, avoiding ships and boats and their bladelike propellers. These are long-lived animals, but modern life in the ocean doesn't make it easy.

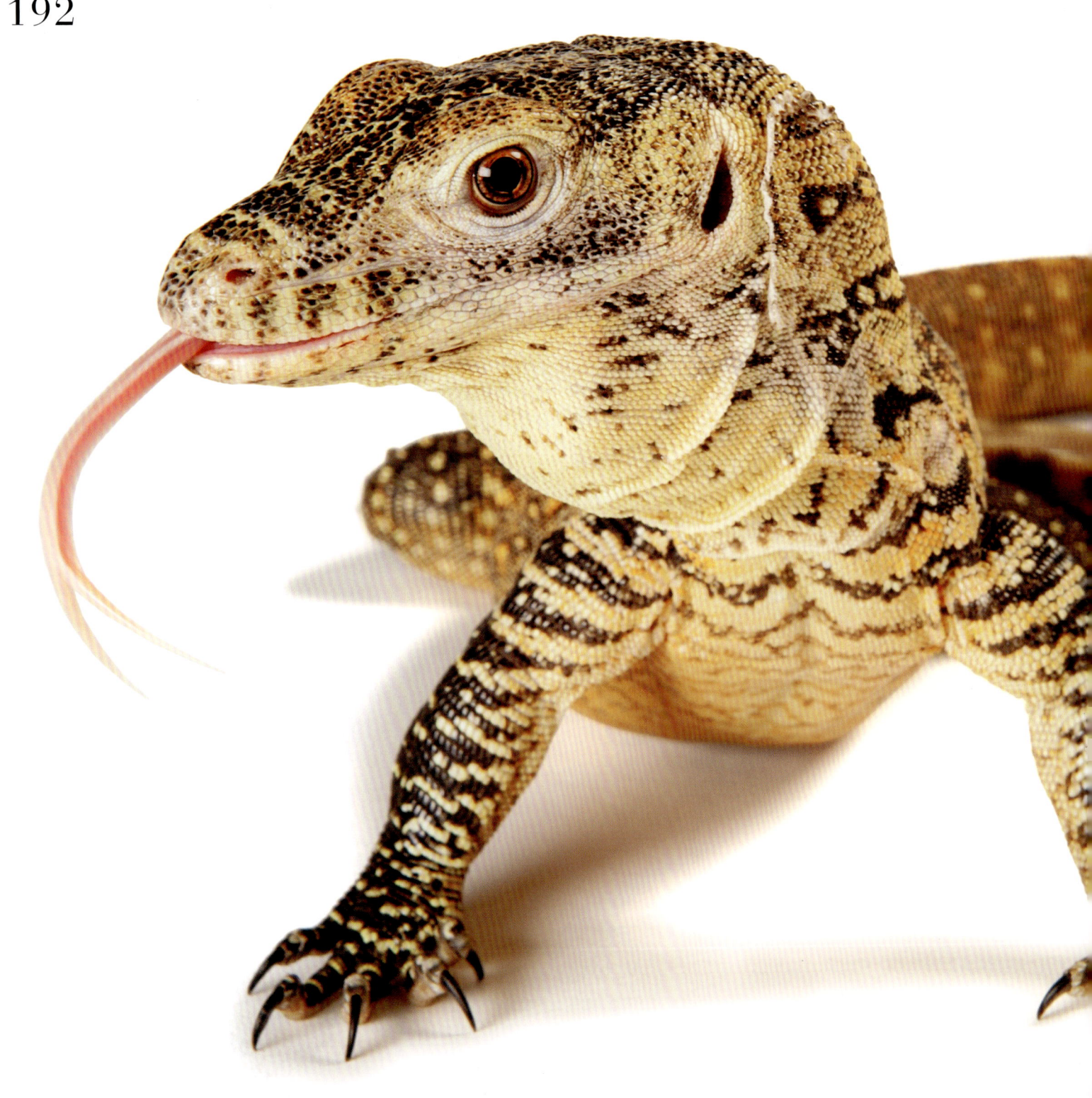

Komodo dragon
Varanus komodoensis (EN)

Komodo dragons are one of only a few dozen species of vertebrates capable of reproduction via mating or parthenogenesis—reproduction without mating, whereby the unfertilized egg develops into an embryo without the need for sperm.

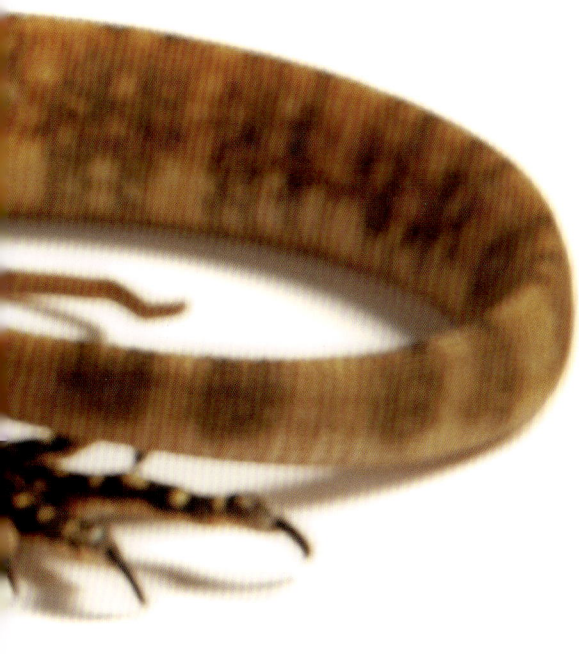

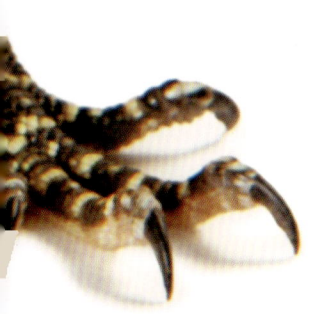

TAKE COVER
Keeping a Distance

T

THE WORLD'S LARGEST living lizard is a hulking, formidable presence on the Indonesian islands where it lives. With its keen senses, serrated teeth, and venomous bite, a Komodo dragon is a fearsome predator. Sometimes the dragon's prey includes its own babies. These cannibalistic tendencies mean young dragons start life with a bang: As soon as they hatch from grapefruit-size eggs, the youngsters scramble into the nearest tree to escape their parents. The young dragons live in trees for about a year, feasting on insects and geckos, before venturing back down to begin life on the ground. Young dragons may also roll in fecal matter, coating themselves with a distasteful smell that adult dragons avoid.

Great horned owl

Bubo virginianus pacificus (LC)

Great horned owls are one of the most formidable hunters in North America, but their fledglings have been known to end up hanging upside down from a tree branch or in a heap on the ground after their first tentative flights.

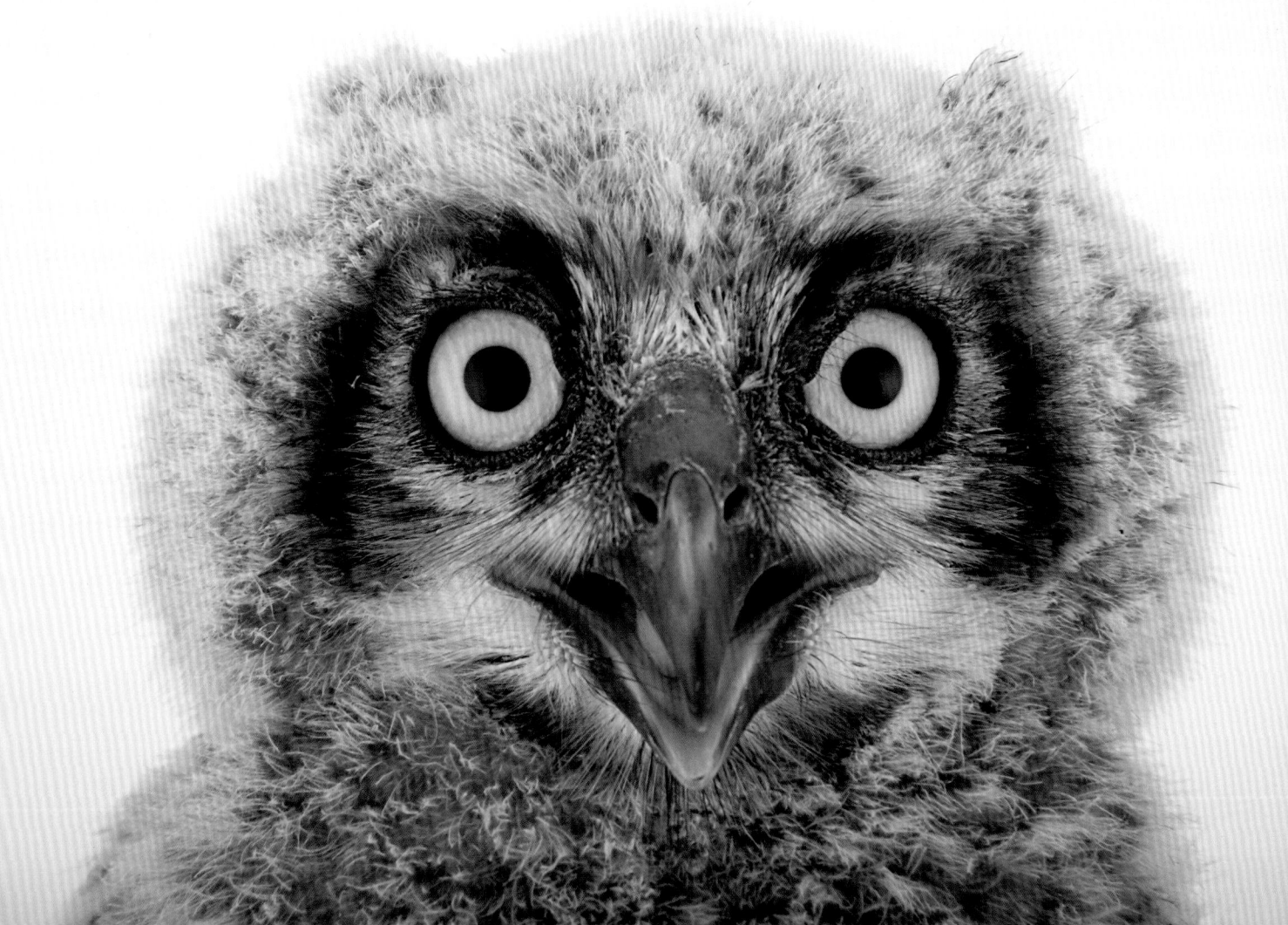

American barn owl

Tyto furcata pratincola (LC)

Barn owlets depend on instinct and practice to learn the skills needed
to hunt successfully. By about 13 weeks, they're ready to spread their wings
in search of their own territory (a process called dispersal).

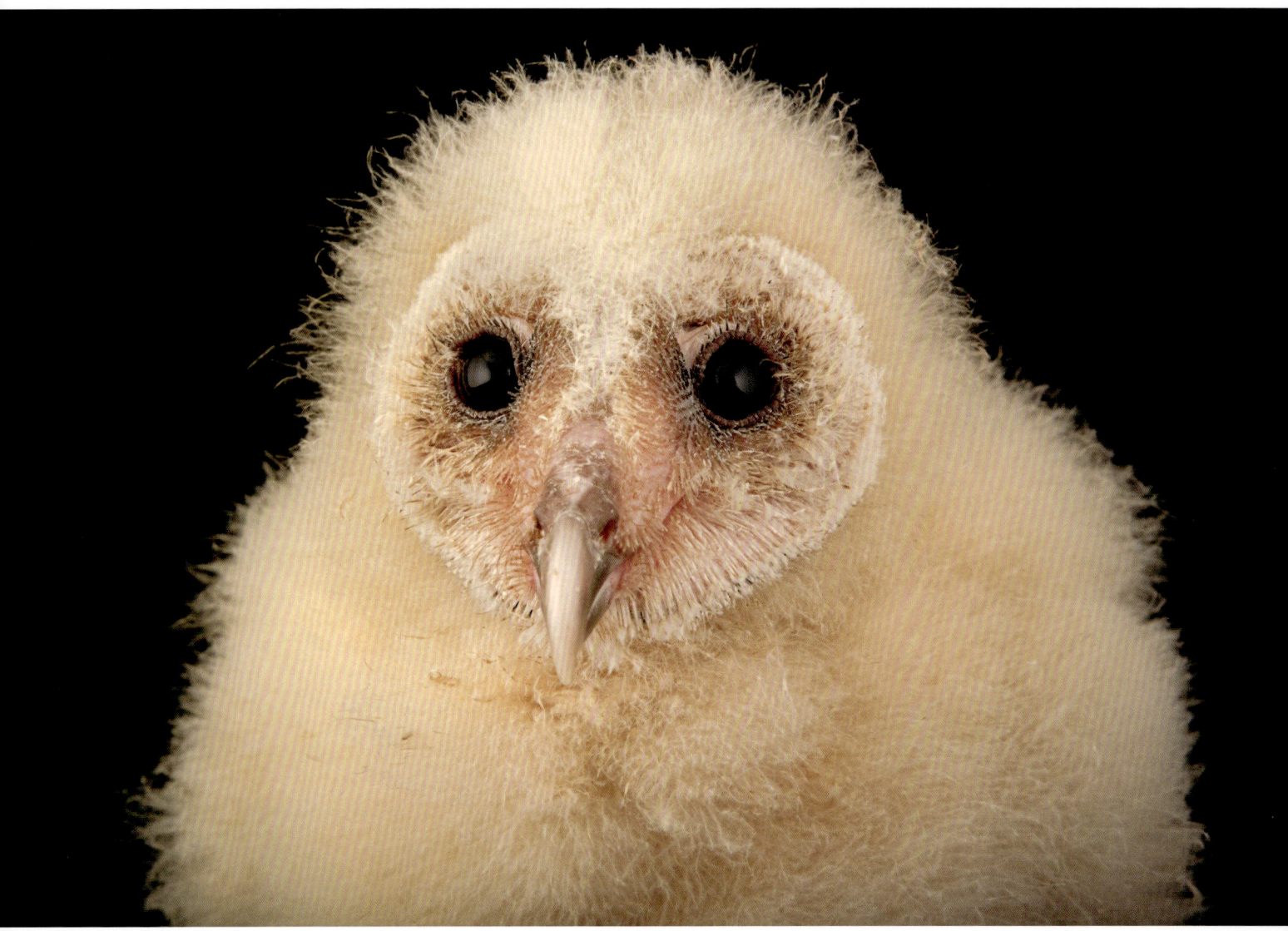

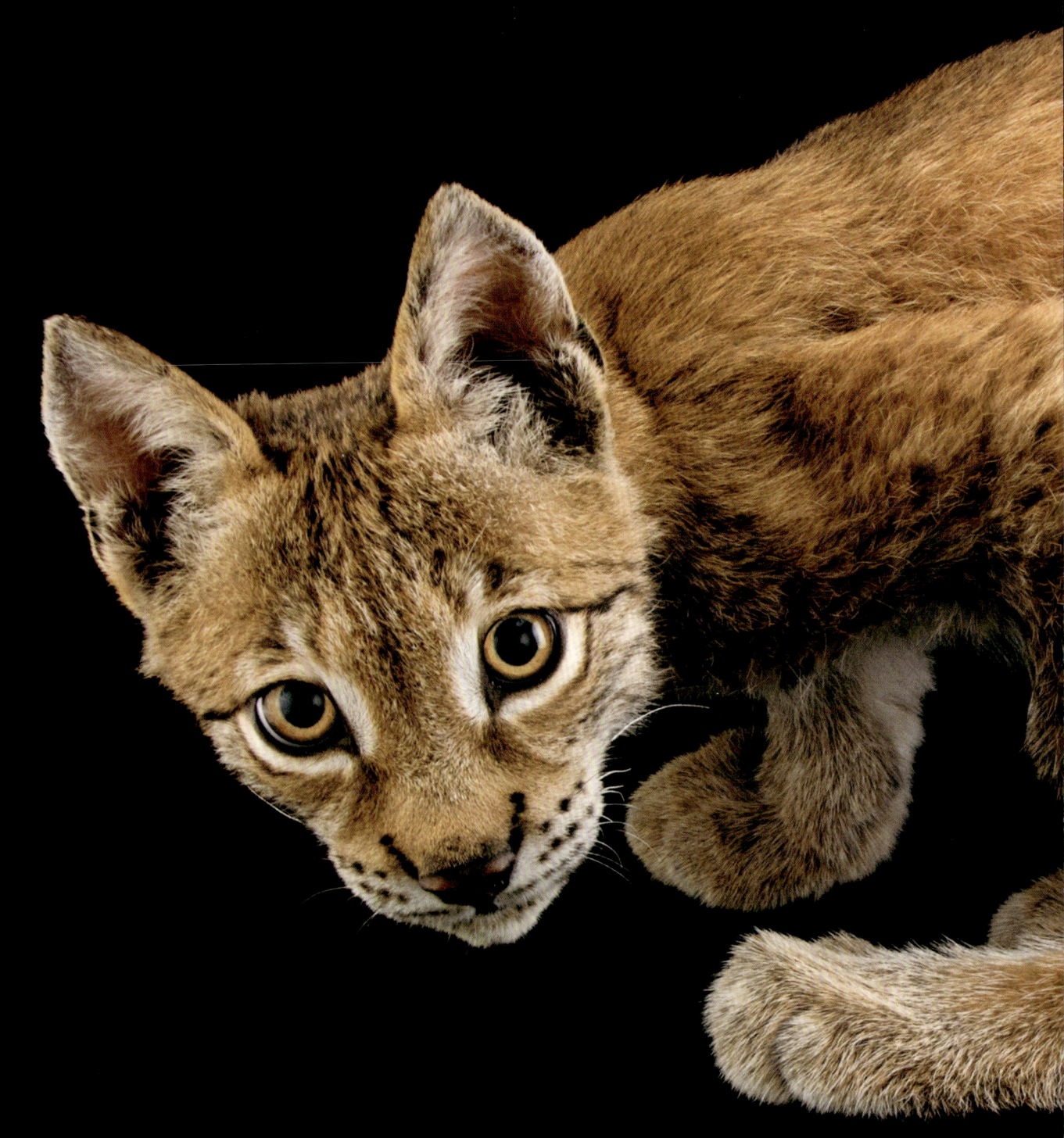

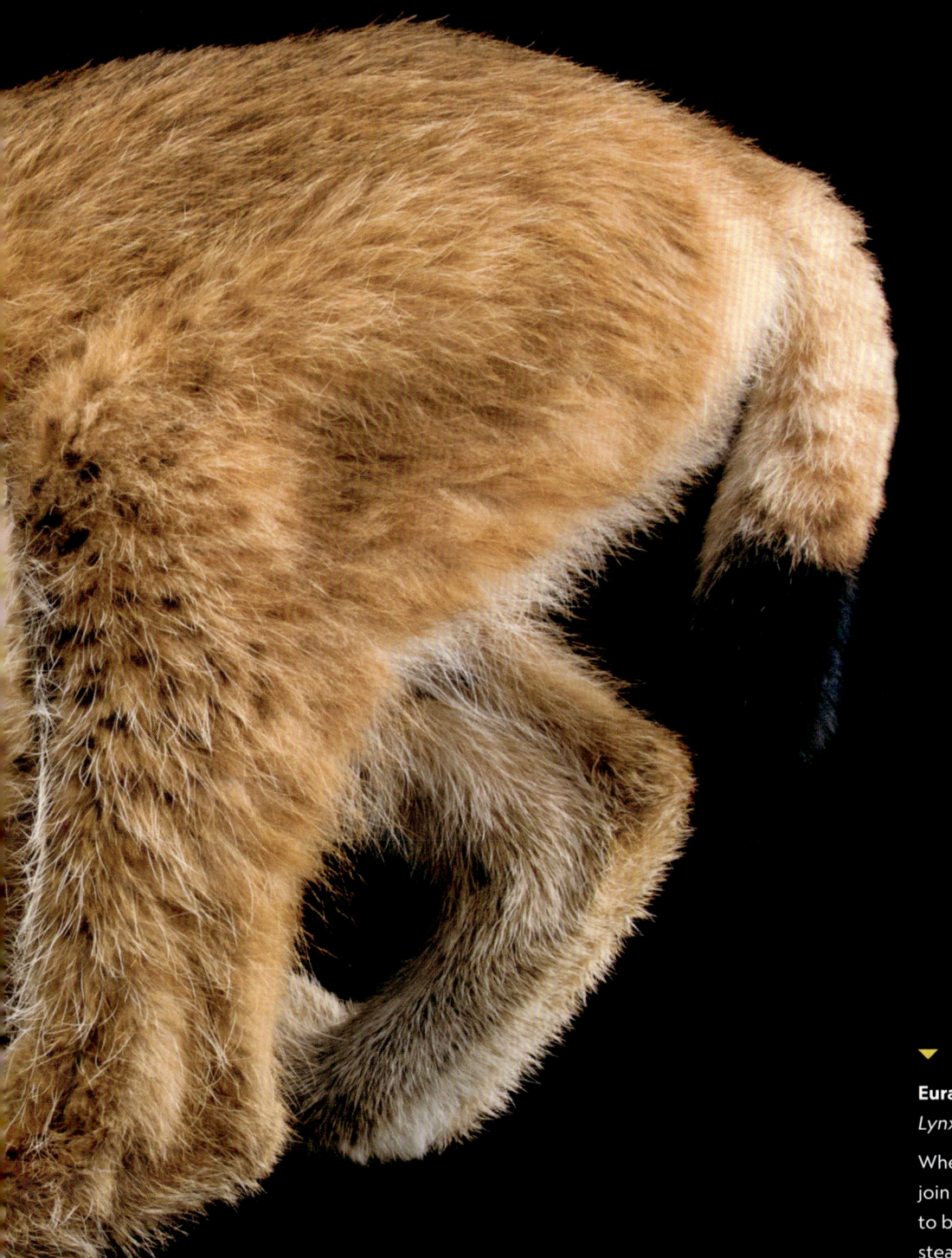

Eurasian lynx

Lynx lynx (LC)

When a lynx kitten is big enough, it will join its mom outside the den to learn how to become a cunning hunter. Lynx use stealth to hunt deer and other prey three or four times their size.

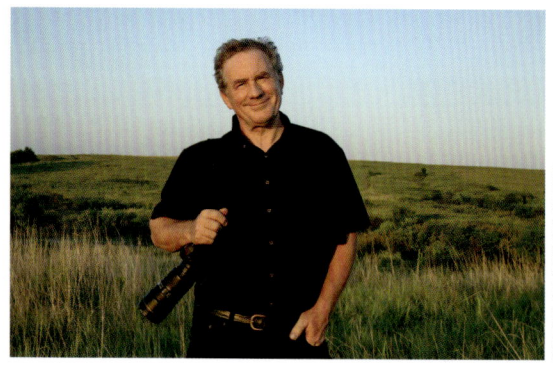

ELLEN SARTORE RUDD

ABOUT THE AUTHOR

JOEL SARTORE is a photographer, author, teacher, conservationist, National Geographic Explorer, and regular contributor to *National Geographic* magazine. His hallmarks are a sense of humor and a Midwestern work ethic. Sartore is the founder of the Photo Ark, a decades-long documentary project to save species and habitats by photographing all animal species in human care around the world. He estimates that the project may total more than 25,000 species, particularly as he has begun to include insects as well. Over the years, Sartore has also contributed to *Audubon* magazine, *Sports Illustrated, Life,* the *New York Times,* and *Smithsonian* magazine. He has published numerous books, including *The Photo Ark, Birds of the Photo Ark, The Photo Ark Vanishing, Photo Ark Wonders,* and *Photo Ark Insects.* Sartore is always happy to return from his travels around the world to his home in Lincoln, Nebraska, where he lives with his wife, Kathy. Their three children and two grandchildren live nearby.

ACKNOWLEDGMENTS

IT TURNS OUT A BOOK like this doesn't happen all by itself.

First, I must search for new animals to photograph. Our Photo Ark taxonomists and scouts Pierre de Chabannes and Nayer Youakim seem to know where everything is and keep me updated. Sometimes they even go along on the shoots, just to make sure I point my camera in the right direction.

Once the trip is done, I drop off a hard drive full of photos and videos at our office in Nebraska, where our team downloads the ones I've marked as favorites. They use Photoshop to clean up the backgrounds, polish captions, add keywords—the works.

Their leader is Rebecca Wright, our office manager. Sarah Booth is our media manager, while Krista Smith keeps our books. Working along with them are Lena Nelson, Forrest Czarnecki, Camille Rotermund, Tanner Bubeck, Dylan Widger, and Jack Mahoney.

We have crews working off-site as well. These folks include Maura Mulvihill, my longtime friend and agent. Renaissance man and friend Clark DeVries ensures our equipment and office building are always in top shape. There's also our social media manager, Taylor Rhoades, and video editor, Alex Crisp. Pierre and Nayer usually rejoin at that point, making sure all the taxonomy is correct before everything gets carved in stone.

Back east at National Geographic headquarters are Colby Bishop and Lindsay Anderson, who manage all things Photo Ark from Washington, D.C. The National Geographic Books team gets it all on the page, including Lisa Thomas, Susan Tyler Hitchcock, Ashley Leath, Meredith Wilcox, Sanaa Akkach, TJ Tucker, Becca Saltzman, Margo Rosenbaum, and more.

And at home are my family, long-suffering from all things Photo Ark at home in Nebraska and around the world, every day of the year: my wife, Kathy; my kids, Cole, Ellen, and Spencer. And we have three new additions to our family now: Jackson Rudd and grandchildren Ezra and Cassidy.

Then there's the engine that drives everything: financing. Kara Ramirez Mullins leads National Geographic's advancement team and always finds the support we need.

And above all, a special thanks to National Geographic CEO Jill Tiefenthaler, who signs off on our grant each and every year. Her vision and generosity keep the Photo Ark moving forward.

My deepest thanks to all.

CONTRIBUTORS

LIBBY SANDER (WRITER) is a journalist, writer, and editor who has been writing for National Geographic Books since 2018. This is her third Photo Ark book. Other recent writing projects for National Geographic include *Secrets of the Whales* and *Sharks: Rulers of the Deep*. A former newspaper reporter for more than a decade, she has also written for the *New York Times,* the *Washington Post,* and the *Chronicle of Higher Education.* Her favorite animal is whichever one she's currently writing about.

PIERRE DE CHABANNES (RESEARCH CONSULTANT) is a photojournalist, explorer, educator, and lecturer. Chabannes has written for *Avicultural Magazine* (Avicultural Society), *Watchbird* (American Federation of Aviculture), and *International Zoo News.* He is a field expedition producer as well as scientific adviser in zoology and conservation and has been an adviser for the National Geographic Photo Ark since 2015.

ABOUT THE PHOTO ARK

THE INTERACTION BETWEEN animals and their environments is critical for maintaining the health of the planet that we call home. But for many species, time is running out. When you remove one, it affects us all. The National Geographic Photo Ark is a multiyear effort to raise awareness of and find solutions to some of the most pressing issues affecting wildlife and their habitats. Founded by Joel Sartore—National Geographic Explorer, photographer, and 2018 Rolex National Geographic Explorer of the Year—the project aims to document every species living in the world's zoos, aquariums, and wildlife sanctuaries; to inspire action through education; and to help protect wildlife by supporting on-the-ground conservation efforts.

Sartore started the National Geographic Photo Ark in his hometown of Lincoln, Nebraska, more than 18 years ago. Since then, he has traveled the world in his quest to create a photo archive of global biodiversity that will feature portraits of more than 25,000 species of birds, fish, mammals, reptiles, amphibians, and invertebrates. Once completed, the Photo Ark will serve as an important record of each animal's existence and a powerful testament to the importance of protecting every one of them.

Join us, and help make a difference by sharing, supporting, and saving. Donate at *natgeo.org/photoark*.

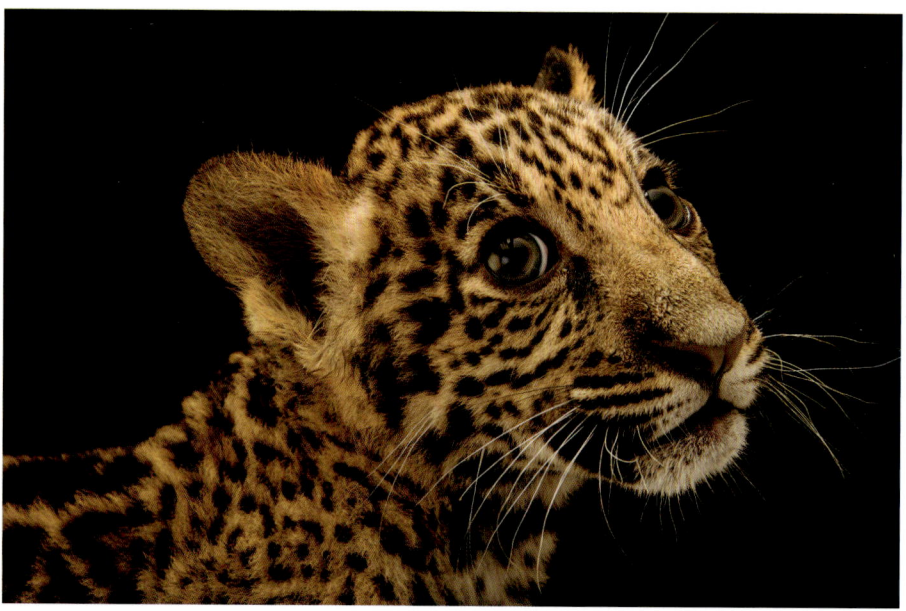

A two-month-old jaguar cub at the Parque Zoológico Nacional in Santo Domingo, Dominican Republic

HOW OUR WORK GETS DONE

I'M OFTEN ASKED EXACTLY how we build the Photo Ark. I'll let the photos do the talking, but generally speaking, it's a multistep process involving a lot of people.

It starts with me going to zoos, aquariums, wildlife sanctuaries, rehabbers, and private collections. Before I get there, we've all agreed on which species I'll be able to photograph.

Small critters—anything that will easily fit in a space two feet by two feet by 40 inches (0.6 m by 0.6 m by 102 cm), from lizards and frogs to birds and rodents—go in my tabletop shooting tent. For big birds, we use a tent so big you could sit in it.

For midsize, toothy mammals, we often use a folding wire kennel purchased at a pet shop.

Big animals are trickier; the zoo often has to paint an off-exhibit room white or black and get the animals used to eating or sleeping inside it, then I come by a week or so later. That way, when we bring the tiger or the bison into the space, he just thinks he's coming for lunch.

For aquatics, we use a series of clear plastic tanks. Thankfully, with no size comparison possible on plain backgrounds, small fish look a lot like big fish.

I still haven't figured out how to light a great whale, by the way.

In the end, the goal is simple: Show animals in their prime, looking healthy and beautiful, so the public will finally realize that each animal is a living work of art, and that every one matters.

Aren't they amazing? Aren't they all worth saving?

You bet they are.

Left: Joel Sartore photographs for the Photo Ark in San Josecito, Costa Rica.
Right: Two juvenile common opossums explore a lens at a wildlife rehabilitation center in Brazil.

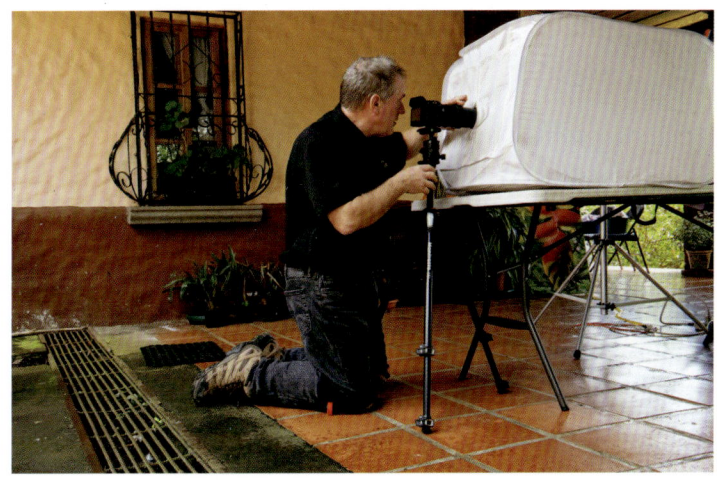
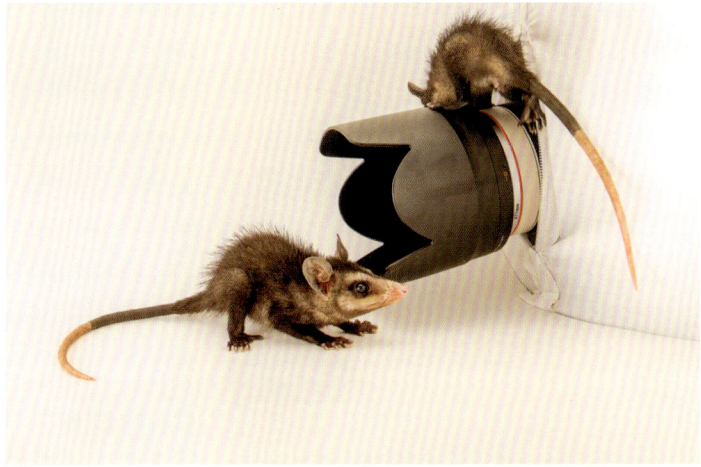

GALLERY KEY
(PAGES 20–21)

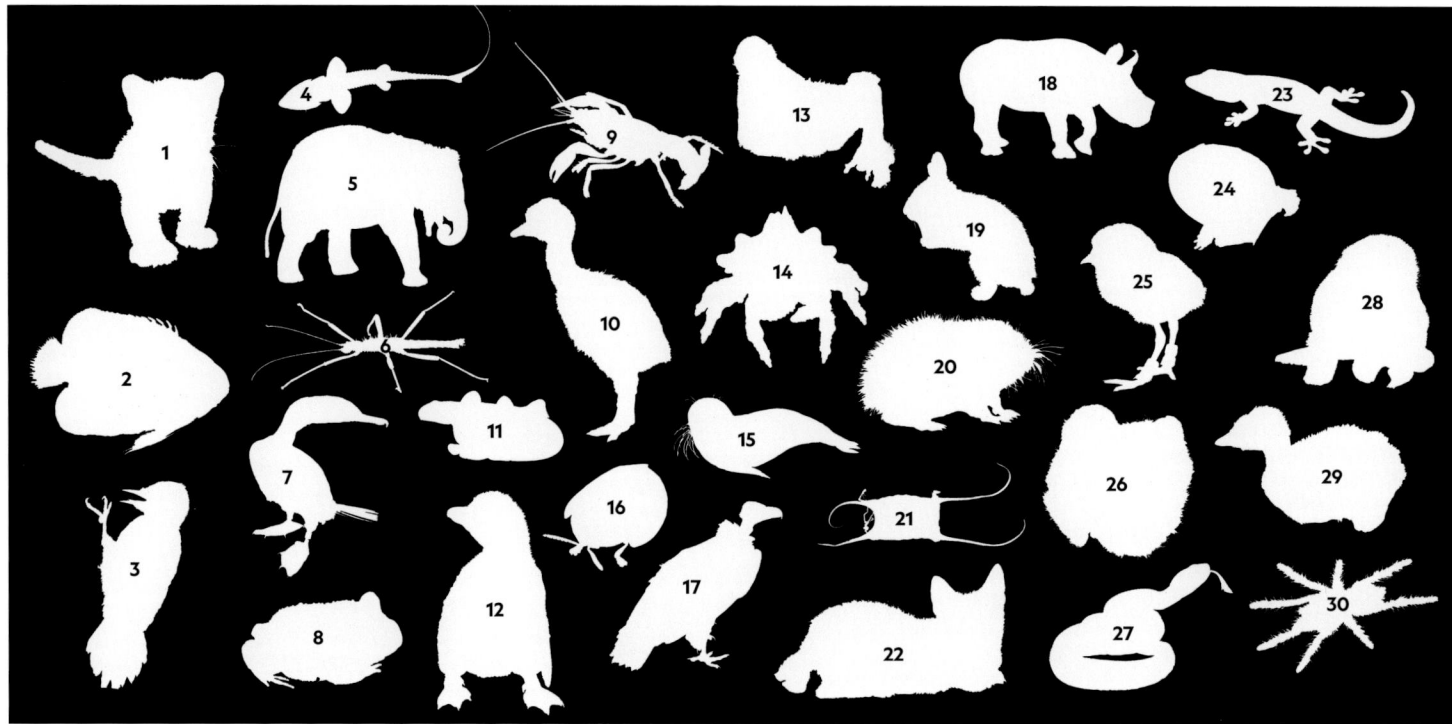

1. Clouded leopard, Columbus Zoo and Aquarium, Powell, OH, *columbuszoo.org.* **2. Emperor angelfish,** Newport Aquarium, Newport, KY, *newportaquarium.com.* **3. Downy woodpecker,** Chintimini Wildlife Center, Corvallis, OR, *chintiminiwildlife.org.* **4. Shovelnose sturgeon,** Missouri River near Atchinson, KS, Missouri Department of Conservation, *mdc.mo.gov.* **5. Indian elephant,** Singapore Zoo, Singapore, *mandai.com/en/singapore-zoo.* **6. Giant stick insect,** Moscow Zoo, Moscow, Russia, *new.moscowzoo.ru.* **7. Pelagic cormorant,** International Bird Rescue, San Pedro, CA, *birdrescue.org.* **8. Stolzmann's horned frog,** Jambatu Amphibian Research and Conservation Center, Pilalo, Ecuador, *anfibiosecuador.ec.* **9. Linear cobalt crayfish,** East Fork Cane Creek, Lewis County, TN. **10. Darwin's rhea,** Lundi Farm, Verl, Germany, *lundi-germany.de.* **11. Papuan epaulette shark,** Loveland Living Planet Aquarium, Draper, UT, *livingplanetaquarium.org.* **12. Humboldt penguin,** Le Parc des Oiseaux, Villars-les-Dombes, France, *parcdesoiseaux.com.* **13. Peruvian spider monkey,** Rainforest Awareness Rescue Education Center, Iquitos, Peru, *rarec.org.* **14. Puget Sound king crab,** Alaska SeaLife Center, Seward, AK, *alaskasealife.org.* **15. Mediterranean monk seal,** Attica Zoological Park, Spata, Greece, *atticapark.com.* **16. Geiger tortoise beetle,** Pine Rocklands, FL. **17. Red-headed vulture,** Parco Natura Viva, Bussolengo, Italy, *parconaturaviva.it.* **18. Eastern black rhinoceros,** Zoo Atlanta, Atlanta, GA, *zooatlanta.org.* **19. Eastern cottontail rabbit,** Carolina Wildlife Center, Columbia, SC, *carolinawildlife.org.* **20. Muskrat,** Wildlife Rehabilitation Center of Minnesota, Roseville, MN, *wrcmn.org.* **21. Clearnose skate egg cases,** Sedge Island Natural Resource Education Center, Barnegat Bay, NJ, *dep.nj.gov/njfw/education.* **22. Andean fox,** Buin Zoo, Buin, Chile, *buinzoo.cl.* **23. Standing's day gecko,** Plzeň Zoo, Plzeň, Czechia, *zooplzen.cz.* **24. Chinese box turtle,** private collection. **25. Western snowy plover,** Monterey Bay Aquarium, Monterey, CA, *montereybayaquarium.org.* **26. Brown-throated sloth,** Toucan Rescue Ranch, San Josecito, Costa Rica, *toucanrescueranch.org.* **27. Florida cottonmouth,** Archbold Biological Station, Venus, FL, *archbold-station.org.* **28. Northern sea otter,** SeaWorld San Diego, San Diego, CA, *seaworld.com/san-diego.* **29. Little grebe,** Hessilhead Wildlife Rescue, Beith, Scotland, *hessilheadwildlife.org.uk.* **30. Trinidad olive tarantula,** Moscow Zoo, Moscow, Russia, *new.moscowzoo.ru.*

ANIMAL INDEX

3 | YOUTH

5 | FLEDGING

IUCN DESIGNATIONS

THE INTERNATIONAL UNION FOR CONSERVATION OF NATURE (IUCN) IS A GLOBAL GROUP DEDICATED TO PROTECTING BIODIVERSITY.
The IUCN Red List of Threatened Species is a comprehensive collection of animal and plant species that have
been analyzed according to their risk of extinction. Once evaluated, a species is placed into one of several categories.
Throughout this book, each species' current IUCN status appears alongside its common and scientific names.

EX | EXTINCT There is no reasonable doubt that the last individual of a species has died.

EW | EXTINCT IN THE WILD Individuals of a species are known only to survive in captivity or as a naturalized population, well outside the former range.

CR | CRITICALLY ENDANGERED The best available evidence indicates that a species faces an extremely high risk of extinction in the wild.

EN | ENDANGERED The best available evidence indicates that a species faces a very high risk of extinction in the wild.

VU | VULNERABLE The best available evidence indicates that a species faces a high risk of extinction in the wild.

NT | NEAR THREATENED A species has been evaluated and does not currently qualify for one of the previous categories but is close to qualifying.

LC | LEAST CONCERN A species has been evaluated but does not qualify for any of these categories.

DD | DATA DEFICIENT Not enough information is available to fully evaluate a species for its risk of extinction.

NE | NOT EVALUATED A species has not yet been evaluated for its risk of extinction.

Since 1888, the National Geographic Society has funded more than 14,000 research, conservation, education, and storytelling projects around the world. National Geographic Partners distributes a portion of the funds it receives from your purchase to National Geographic Society to support programs including the conservation of animals and their habitats.

National Geographic Partners, LLC
1145 17th Street NW
Washington, DC 20036-4688 USA

Get closer to National Geographic Explorers and photographers, and connect with our global community. Join us today at nationalgeographic.org/joinus

For rights or permissions inquiries, please contact National Geographic Books Subsidiary Rights: bookrights@natgeo.com

Photographs and Introduction Copyright © 2025 Joel Sartore. Compilation Copyright © 2025 National Geographic Partners, LLC. All rights reserved. Reproduction of the whole or any part of the contents without written permission from the publisher is prohibited.

NATIONAL GEOGRAPHIC and Yellow Border Design are trademarks of the National Geographic Society, used under license.

Library of Congress Cataloging-in-Publication Data
Names: Sartore, Joel, author. | National Geographic Society (U.S.)
Title: National Geographic Photo Ark babies / Joel Sartore.
Other titles: Photo Ark babies
Description: Washington, D.C. : National Geographic, [2025] | Includes index. | Summary: "This adorable collection in Joel Sartore's Photo Ark series highlights youngsters in the animal kingdom"-- Provided by publisher.
Identifiers: LCCN 2024017085 | ISBN 9781426224065 (hardback)
Subjects: LCSH: Wildlife photography. | Animals--Infancy--Pictorial works. | Zoo animals--Pictorial works. | LCGFT: Illustrated works.
Classification: LCC TR729.W54 S25 2025 | DDC 779/.32--dc23/eng/20240725
LC record available at https://lccn.loc.gov/2024017085

Printed in China

24/RRDH/1

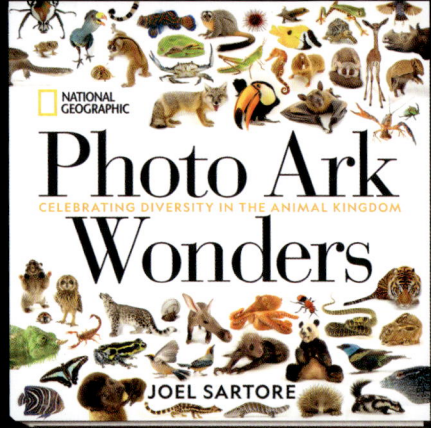

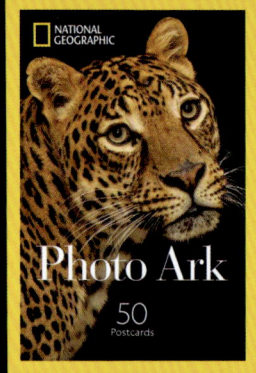

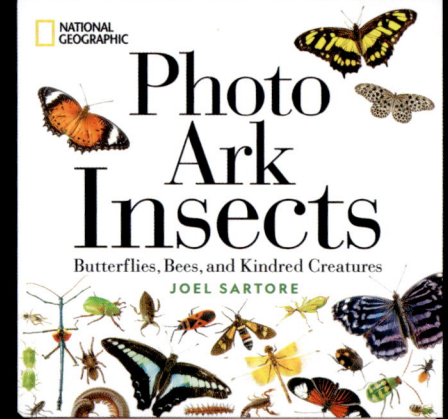

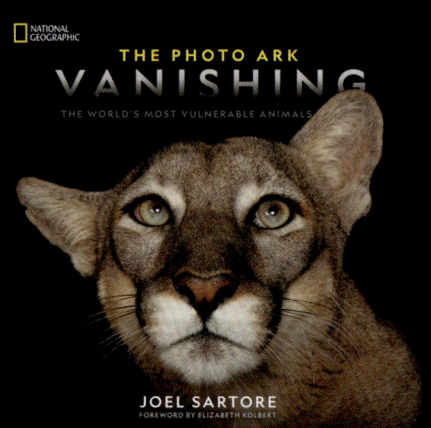

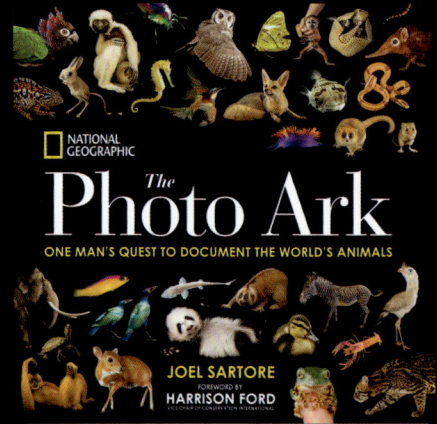

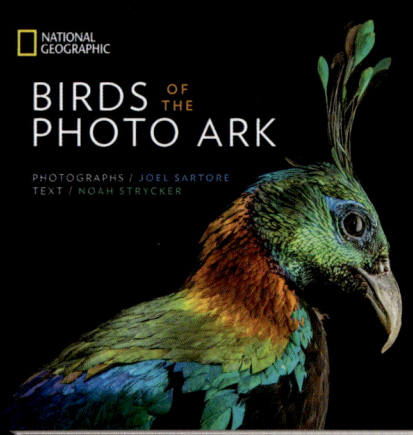

AVAILABLE WHEREVER BOOKS ARE SOLD

@NatGeoBooks

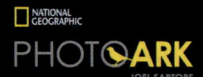

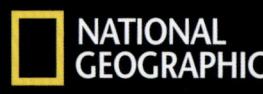

© 2024 National Geographic Partners, LLC